Aquaculture Machineries and Equipments

NIPA® GENX ELECTRONIC RESOURCES & SOLUTIONS P. LTD.
New Delhi-110 034

About the Authors

Dr. Mohammad Tanveer did his B.Tech. in Agricultural Engineering from Chandra Shekhar Azad University of Agriculture and Technology, Kanpur (Uttar Pradesh), M.Tech. and Ph.D. in Aquacultural Engineering from IIT Kharagpur.

After his M. Tech. in 2007 he joined as an Aquaculture Engineer in a private firm followed by PhD from IIT Kharagpur. In 2013 he visited Haramaya University, Ethiopia as a Senior Lecturer.

In 2015 September, Dr. Tanveer joined Tamil Nadu Dr. J. Jayalalithaa Fisheries University, Nagapattinam as Assistant Professor in College of Fisheries Engineering. At Present Dr. Tanveer is working as Associate Professor in College of Fisheries Engineering, Tamil Nadu Dr. J. Jayalalithaa Fisheries University, Nagapattinam.

During his more than 7 years of career, Dr. Tanveer contributed for his profession by writing Two books, Aquacultural Engineering Manuals, conducted more than 10 Training programs and prepared Manuals in College of Fisheries Engineering, published 20 research and review papers in international and national journals, published more than 10 number of popular articles, delivered talks on TV and radio programmes, attended more than 12 national and international conferences. He completed one Research Project funded by the University. Currently, He is handling one number of externally funded research project as a Co-PI.

Dr. Tanveer has been the member of the World Aquaculture Society, Indian Society of Agricultural Engineers and many national professional societies. He is connected with many shrimp and fish farmers for the betterment of the sector.

Er. Pradeep Ramesh, received B.Tech Fisheries Engineering in the year 2019 and M.Tech in Aquacultural Engineering degrees in 2021 from Tamil Nadu Dr. J. Jayalalithaa Fisheries University (TNJFU), Nagapattinam TN, India. Presently he is working as Assistant Professor (C) in Department of Aquacultural Engineering College of Fisheries Engineering, TNJFU, Nagapattinam TN, India. He has qualified ICAR- NET in Agricultural Structures and Environmental Management during 2021.

He has qualified ICAR-ARS Mains. In a very starting stage of his career and he has published 2 research papers of International repute and many more are under review and 4 numbers of Technical articles. Moreover, he is also involved in writing a book on Aquaculture machineries and equipments. He has attended 10 numbers of International conferences/Seminar and Webinars.

Er. S. Ayesha Jasmin earned her B.Tech Fisheries Engineering degree from Tamil Nadu Dr. J. Jayalalithaa Fisheries University (TNJFU), Nagapattinam TN, India in the year 2022. She has proven expertise and publication records in high impact journals related to application of Artificial Intelligence in Aquaculture. She is presently working as Lab Technician (C) in Central Instrumentation facility at College of Fisheries Engineering, Nagapattinam

Aquaculture Machineries and Equipments

M. Mohammad Tanveer
Associate Professor &Head
Department of Aquacultural Engineering
College of Fisheries Engineering
Tamil Nadu Dr. J. Jayalalithaa Fisheries University
Nagapattinam-611002

Er. R. Pradeep
Assistant Professor (Contractual)
Department of Aquacultural Engineering
College of Fisheries Engineering
Tamil Nadu Dr. J. Jayalalithaa Fisheries University
Nagapattinam-611002

Er. S. Ayesha Jasmin
B.Tech (Fisheries Engineering)
Department of Aquacultural Engineering
College of Fisheries Engineering
Tamil Nadu Dr. J. Jayalalithaa Fisheries University
Nagapattinam-611002

NIPA® GENX ELECTRONIC RESOURCES & SOLUTIONS P. LTD.
New Delhi-110 034

NIPA® GENX ELECTRONIC RESOURCES & SOLUTIONS P. LTD.

101,103, Vikas Surya Plaza, CU Block
L.S.C.Market, Pitam Pura, New Delhi-110 034
Ph : +91 11 27341616, 27341717, 27341718
E-mail:newindiapublishingagency@gmail.com
www: www.nipabooks.com

For customer assistance, please contact
Phone: + 91-11-27 34 17 17
Fax: + 91-11- 27 34 16 16
E-Mail: feedbacks@nipabooks.com

ISBN: 978-81-19002-41-2

Composed and Designed by NIPA.

Preface

This text book entitled "Aquaculture Machineries and Equipments" has been prepared to meet the requirement of undergraduate and post graduate students of Fisheries Engineering/Aquacultural Engineering. There is no specific book available for this subject to cover many of the topics. Therefore, interested persons and students are feeling worried to have a basic idea to different machineries and equipment's used and also its optimized usage conditions. The design and management of these machineries and systems is based on their production goals and the economics of the farming operation. This book is answer to this requirement.

The idea of preparing book on 'Aquaculture Machineries and Equipment's' came from the suggestions made by students of B.F.Sc., B.Tech (Fisheries Engineering) and others who are interested. Thus, the needs of relevant topics were discussed explicitly using suitable examples. The book can also be used by researchers and persons involved in developing machineries in the field of Aquaculture Engineering.

Authors

Contents

Contents

1

Selection of Pumps

1.1 Introduction

Pump is a mechanical device to increase the pressure energy of fluid. Aquacultural enterprises often find it necessary to move fluid against an energy gradient. Most commo n may be filling or emptying of tanks. For performing an assigned job, the best selection of pumps should be such that it should work at its peak efficiency. Unfortunately, there is only one head and discharge condition for each pump where the peak efficiency is obtained and therefore, it is rarely possible to achieve the highest efficiency for a pumping condition. Manufacturers cannot design and build so many varieties of pumps to meet all operating conditions. They have therefore, settled upon some standard designs which are generally required for most of the head and discharge conditions. It may be possible to select some pump from the series of pumps available in the market and obtain efficiency slightly lower than the peak value.

Before we go for selection of pumps, some of the following terminologies are required to be known.

1.2 Head of a pump

i) Static head,

ii) Manometric head,

iii) Total or gross head.

Static head: It is the vertical distance from the water surface at the source to the point of delivery. It consists of static suction head (H_s) and static delivery head (H_d)

Manometric head: It is the head measured across the pump inlet and. outlet flanges. It expresses the increase in pressure energy per unit weight of liquid handled by the impeller.

$$Hmano = (P_d - P_s)/l + h_g \quad ...(1.1)$$

where P_d = pressure gauge reading (kg/m^2),

P_s = vacuum gauge reading (kg/m^2),

1 = density of water (kg/m^3) and

h_g = vertical distance between the tapping for pressure and vacuum gauges.

Net positive suction head (NPSH): The NPSH is a characteristic of the pump and should be furnished with the pump characteristic curves. For operation of a centrifugal pump without cavitation, the suction lift plus all other losses must be less than the theoretical atmospheric pressure.

Total, gross or effective head (H): This is the actual head against which the pump has to work. It is equal to the summation of manometric head and the velocity head. For all practical purposes, $H = H_{mano}$

1.3 Loss of head in pipe line and fittings

The different points where head losses occur in a pumping system are given below:

i) Footvalve and strainer,

ii) Friction loss through suction and delivery pipes,

iii) Head loss in bends,

iv) Head loss in reducers,

v) Head loss in valves,

vi) Exit velocity head loss.

In case the pump is used to lift water from tubewells, then aquifer and screen losses occur in place of footvalve and strainer losses.

All head losses from the foot valve upto the outlet of the delivery pipe is given in the equation (3)

$$H_L = H_{LS} + H_{Ld} + \sum H_{Lf} + \frac{V^2}{2g} \quad \dots (1.2)$$

Where, H_L total head loss,

H_{Ls} frictional loss in suction pipe $= \dfrac{fl_s v^2}{2gd}$.

H_{Ld} frictional loss in delivery pipe $= \dfrac{fl_d v^2}{2gd}$, f = frictional factor,

l_s = length of suction pipe,

l_d = length of delivery pipe,

v = exit velocity and

d = diameter of pipe.

$\frac{v^2}{2g}$ = loss of kinetic head due to delivery of water at exit,

$\sum H_{Lf}$ = Sum of head losses in pipe fittings, it includes strainer, foot valve, bends, regulating and non-return valve and all changes of cross section.

1.4 Efficiency of pump

It is the ratio of the power supplied by the pump to the power delivered to the pump shaft.

η = output/input = (γ Q H mano)/[75 (BHP-HP lost in coupling mechanism)]

BHP is the brake horse power of the electric motor or prime mover.

1.5 Selecting a Pump

For a particular job the best selection of a pump will be one that will operate at its peak efficiency. Unfortunately, this is rarely possible for there is only one capacity and one head condition for each pump where the highest efficiency is obtained. It is obviously impossible for any manufacturer to design and build many pumps required to meet the operating conditions. Manufactures have, therefore, settled upon standard design for required head and capacity ranges. A well designed and integrated line of pumps will be so arranged that it is possible to select some pump from the line for any condition and obtain an efficiency that is within a few percentages point of the maximum. Characteristics curves are available and should be used to select the best pump for the particular job.

The characteristics curves relate the major pump parameters of head, efficiency, discharge and power requirements. Each pump has a unique set of these chrematistics and will in fact have different characteristics after it has operated for a period of time. In spite of the difference in performance characteristics between individual pumps and the changes with time, pump with similar construction display similar characteristics curves. For new pumps, the differences between the performance characteristics of individual pumps of the same design are insignificant. The major effect of wear is rapid reduction of pump efficiency and reduced head and/or pressure capability.

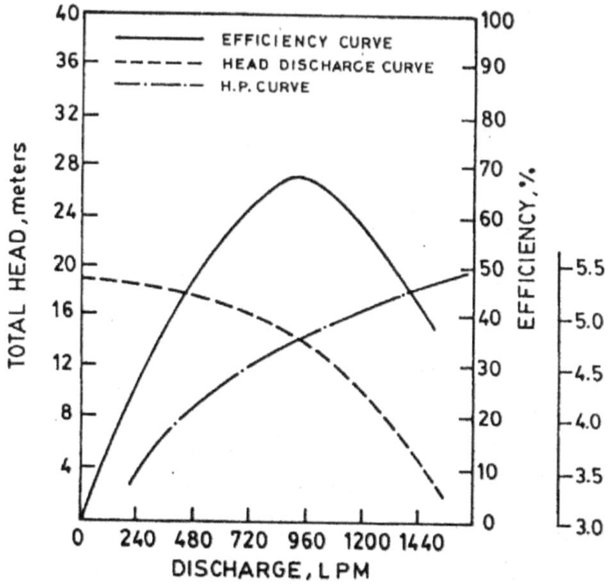

Fig. 1.1: Characteristics curves of a centrifugal pump

The head-discharge curve relates the maximum head produced at any discharge for the specified pump operating at constant speed. For centrifugal pumps, discharge decreases as the head requirement increases. The efficiency curve specifies pump efficiency for each operating condition. Operating costs will be minimized by operating at maximum efficiency. Maximum power requirement occurs at a discharge some what above discharge for maximum efficiency. An important point about centrifugal pump characteristics is that power requirements are reduced as the discharge is reduced from the point of maximum efficiency. This allows the restricting the discharge without risk of overloading the power source, but the efficiency is reduced. Thus only in very special circumstances in aquacultural enterprises the over sized pump may be selected and discharge regulated by choking them.

The pump characteristics change by changing the speed. In a belt driven pump, it is relatively easy to change the speed by changing the pulley size. Therefore, characteristics curves relate head, discharge, efficiency, pump speed, impeller diameter and power requirements. Any one characteristics curve may show the relationship of upto 5 of these variables with the sixth held constant. The set of variables usually depend on the expected use of the pump.

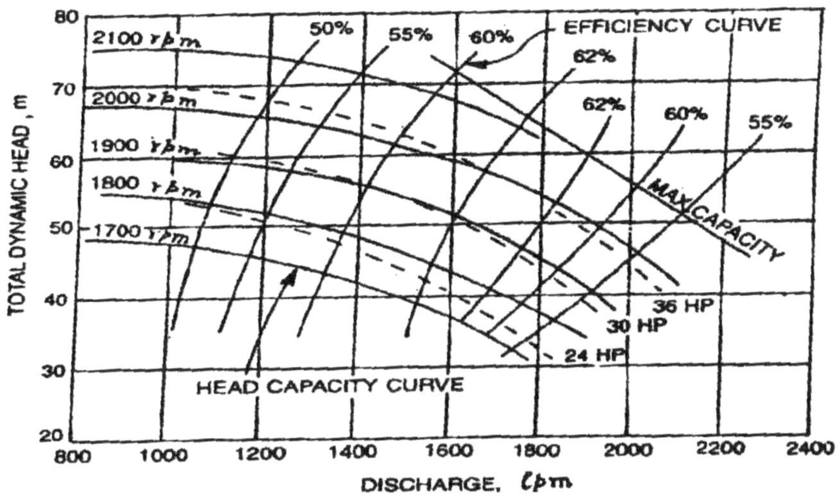

Fig. 1.2: Typical characteristics curves for horizontal centrifugal pump with variable speed

1.6 Propeller Pumps

In most of the aquacultural farms lifting of water involves low head and high discharge, especially, true for brackish water aquaculture. For such requirements also centrifugal pumps are used. Analysis of the characteristics curves of centrifugal pumps revealed that they should not be used for a total head of less than 4 meter as their efficiency for this range of total head fells below reasonable value. In these situations, the propeller pump is the best suited. It has relatively high discharge capacity and high efficiency compared to other pumps for lifting water against low heads from one to four metres. Mixed flow pumps are used for medium heads in cases where the desired capacity is above the range of the centrifugal pumps and below that of the propeller pumps.

Unfortunately, propeller pumps have not become popular in India as yet. There are only a few firms in India manufacturing propeller pumps for specified head and discharge ranges and the prices are disproportionately high. Due to this, people prefer to purchase centrifugal pumps at low initial investment but because of low efficiency require high energy input at low heads. This results in high cost of pumping and heavy loss of energy.

Now a cheap type propeller pump has been developed at G.B. Pant University of Agriculture and Technology and it has been named as Pantnagar Propeller pump.

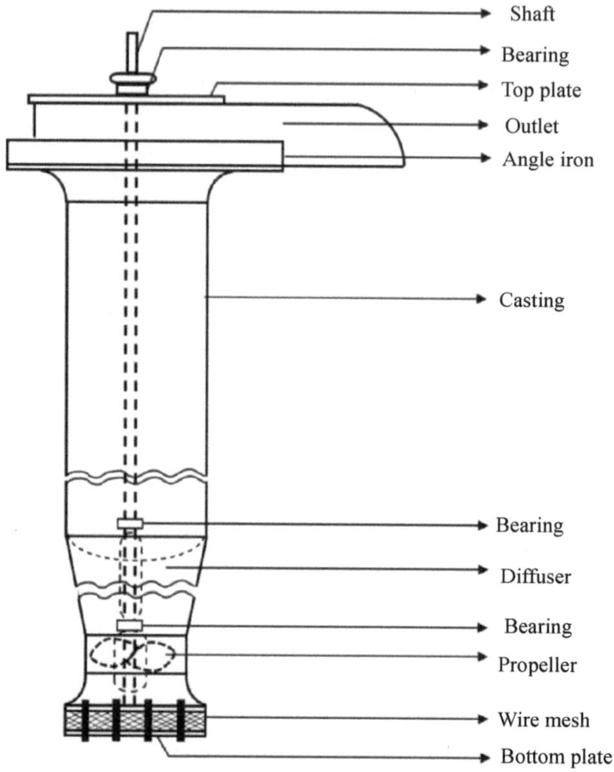

Fig.1.3: Schematic sketch of a Pantnagar propeller pump

The principal parts of a propeller pump are

a) Casing
b) Propeller
c) Diffuser
d) Shaft
e) Supporting end plates

1.6.1 Installation and Operation of Propeller Pump

The propeller pump may be installed in a tank, pond, river, open well or tidal creek. It is mounted on a mild steel channel frame or wooden frame placed on the side of the embankment. The pump can be operated by a 5 H.P. motor or diesel engine of 1440 rpm either by direct drive system or belt-pulley arrangement. The following steps may be followed for installation and operation of a propeller pump.

i) In case of direct drive, the motor is fixed on a mild steel frame in which the pump is supported. The shaft of the pump is connected with the shaft of the motor using a coupler.

ii) For operating the pump at a speed other than motor speed, belt and pulley arrangement is adopted. In this case the motor is mounted on a M.S. channel frame on a concrete foundation adjacent to the pump. The pulley on the pump and drive should be adjusted to get the desired rpm of pump.

iii) There should be a clearance of at least 50 cm between the pump casing and the side of the channel.

iv) The gap between the bottom end plate of the pump and the bottom of the channel stream should be at least 50 cm

v) The pumping water level should be 50 cm above the propeller.

vi) There should not be any submergence at the outlet. Free flow must be maintained.

vii) Before installation of the pump, the shaft may be rotated by hand to ensure that the propeller is not striking the casing and the bearings are properly functioning.

1.7 Air lift pump

It utilises compressed air for raising water. Because the density of a mixture of air in water is much lower than that of pure water, if such a mixture is balanced against a water column, the former will raise much higher that the later.

Compressed air is supplied by a compressor unit installed on the surface. The rapid stream of air at A produces a jet and entraps the water in the immediate vicinity and comes it upward to B. The air is introduced at a considerable depth below the water surface in the well, so that the difference $h_m - h_s = h_1$ may be sufficient to initiate flow to B, the desired delivery level. For best results the lift h_1 should be less than hs.

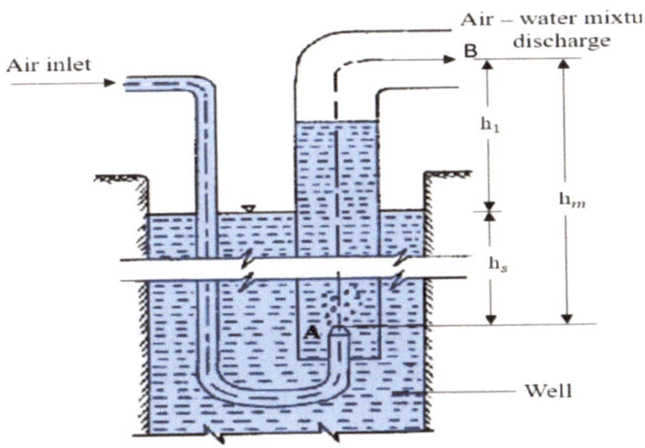

Fig.1.4: Schematic sketch of an air lift pump

Advantages

i) No moving parts, no valves. No damage due to solids in suspension in water.

ii) Raises more water for a given diameter than any other pump.

iii) Can be used to drain out water in mines where the compressor units are already available.

iv) For aquaculture, water with high air content can be supplied.

Disadvantages

i) Efficiency is very low. The volume V of air is in m^3 at atmospheric pressure, required to lift one m^3 of water through height h1 depends on the efficiency.

If, $\eta = 30\%$, $V \cong 1.75 h_1 m^3$, $\eta = 40\%$, $V \cong 1.45 h_1 m^3$

ii) Possibility of air leakage.

<u>Theory of flow</u>

From fluid statistics, the relationship between densities and height of rise are given below.

$$h_m . \gamma m = h_s . \gamma 1 \qquad \ldots (3)$$

where, h_m = height of liquid air mixture in the tube,

h_l = height from free water surface to tube discharge, *i.e.* lift,

γ_m = specific gravity of liquid-air mixture,

h_s = depth of submergence of air inlet to pump and

γ_1 = specific gravity of the liquid outside the tube.

As $\gamma m < \gamma 1$, h_m must be larger than hs to satisfy the above equation.

1.8 Fitting of pumps and wells

The drawdown-discharge characteristics of the well influence the selection of a pump for a particular well. The pump that best suits a well can be selected by matching the characteristics of the pump with that of the well.

1.8.1 Well characteristics

The yield of a well depends on how much the water table is lowered by pumping. The well yield is the volume of water discharged from a well per unit of time. The discharge-drawdown relationship of a well is known as the well characteristics. Data on this relationship may be plotted on a graph paper with level as measured from the ground surface, as the ordinate.

1.8.2 Matching of pump and well characteristics

The characteristic curves of the well and the pump enable the selection of the pump which best suits the well. It is usual to draw the well characteristics curve on a tracing paper in the same scale as that of the pump curve. The tracing paper is placed on the graph showing the pump characteristics. The characteristics of pumps manufactured by leading manufacturers are matched with the well characteristics. This will lead to the selection of the pump which gives the maximum efficiency and the desired discharge at the estimated total head. The matching of the characteristic's curves of a pump with well characteristics is shown in Fig-1.5. The point of intersection of the head capacity curves of the pump and well provides a point of selection. At this point the pump will give a discharge of 3150 liters per minute against a total head of 13 meters and an efficiency of 65 per cent. The power input is 14 horse power. By careful selection from a wide range of products, it is possible to select a pump which will give the desired discharge rate at the estimated total head and provide an efficiency of 75 to 80 per cent or more.

Some pump manufacturers provide information of their products as tabular data. Though precise matching of pump and well characteristics is not possible with tabular data, by checking the data of a large number of pumps, it is possible to select an efficient pump.

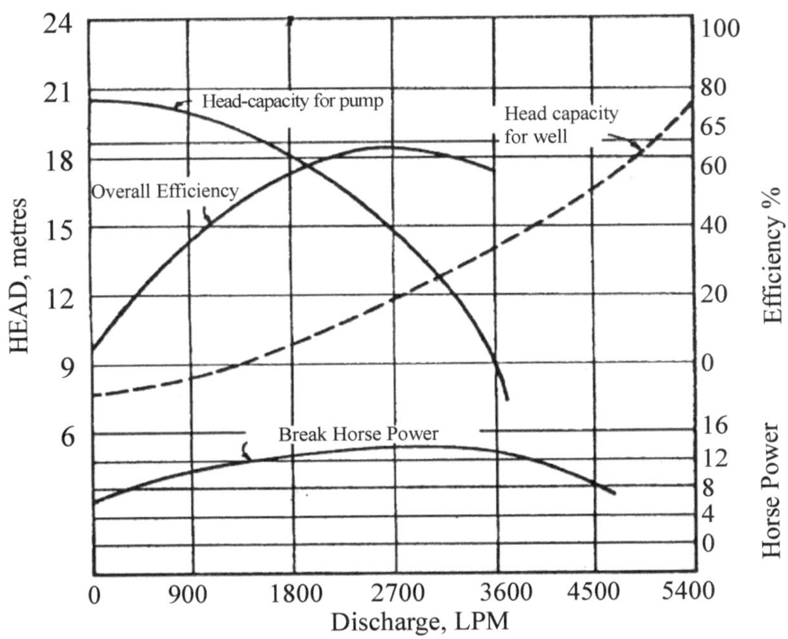

Fig. 1.5: Matching of well and pump characteristics

Example Problems

1.1. Calculate the discharge of reciprocating pump (single acting) if area of cylinder is 0.25 m², length of stroke is 0.15 m, number of cylinder =1 and speed of pump is 50 rpm.

Solution

Discharge Q= A×L×N/60 × number of cylinder× number of acting

A= 0.25 m^2

L= 0.15 m

N= 50 rpm

n = 1

Number of acting =1

$$Q = 0.25 \times 0.15 \times \frac{50}{60} \times 1 \times 1 = 0.03125 \text{ m}^3/\text{s}$$

1.2. Calculate the power required to drive the single acting reciprocating pump for water of following specification.

Solution

Area of cylinder	=	0.85 m^2
Length of stroke	=	0.35 m
Number of cylinder	=	2
Speed of pump	=	15 rpm
Suction head	=	0.5 m
Discharge head	=	1 m

$$Power(hp) = \frac{WALn(h_s + h_d)}{4500}$$

$Where\ w = p_g$

$$Power = \frac{1000 \times 9.81 \times 0.85 \times 0.35 \times 15 \times 2 \times (0.5 + 1)}{4500} = 29.1 h.p.$$

1.3. Calculate water horse power for centrifugal water pump if flow rate is 4500 liter/h. head added to the flow is 10 m.

Solution

1 m³ = 1000 L

$$Q = \frac{4500 \times 10^{-3}}{60 \times 60} m^3 / s$$

$$Water\ horse\ power\ WHP = p_g HQ$$

$$= \frac{1000 \times 9.81 \times 10 \times 4500 \times 10^{-3}}{60 \times 60} = 12.26\,hp$$

1.4.Calculate brake horse power for centrifugal water pump if flow rate is 1000 litres/h. Head added to the flow is 10 m pump efficiency is 86%.

Solution

1 m³ = 1000 L

$$Q = \frac{1000 \times 10^{-3}}{60 \times 60}$$

$$Brake\ horse\ power\ bhp = \frac{p_g HQ}{n}$$

$$bhp = \frac{1000 \times 9.81 \times 10 \times 10000 \times 10^{-3}}{0.86 \times 60 \times 60} = 316.86\,hp$$

1.5. A pump fed aquacultural farm consists of 50 ponds of area 100 m × 40 m each, the average depth of water is 120 cm in the ponds. The ponds require 10% water exchange after an interval of one day. The centrifugal pumps selected for this purpose have to be operated for 10 hours a day. Determine the number of pumps required and their power for this purpose. Select suitable diameter of suction and delivery pipes so that the velocities in the suction and delivery pipes may be about 2 and 3 m/s respectively. Overall efficiency of the pump = 40 %, total static head = 12 m. Distance from the point of pumping to the point of delivery = 20 m. Each delivery pipe should be fitted with a gate valve and one 900 long radius elbow, each suction pipe is fitted with a foot valve and strainer. Kinematic viscosity of water (ν) = 1.054 × 10-6 m2/ s at 200 C temperature. Assume 15% of the water exchange as the evaporation and seepage losses.

Solution

Total volume of water exchanged per day = 100×40×(50/ 2)×1.2×(10/ 100) = 1200 m³/ day

(For 10 hours of operation, Q = 0.3333 m^3/s)

15 % of this is the evaporation and seepage loss

Total discharge = 0.3333 × 1.15 = 0.38333 m^3/s

Velocity in suction pipe (V_s) = Q/A = [0.38333 / n/ 4 $(d_s)^2$]

Taking Vs = 2 m/s; ds = 49.4 cm, similarly for Vd = 3 m/s; dd = 40.33 cm

Suction side

Reynolds Number (Re) = V D/ν = 2 × 0.494/1.054 × 10^{-6} = 9.37 × 10^5

For GI pipe, ∈ = 0.0005 ft = 1.5244 × 10^{-4} m

Relative roughness = ∈ / D = 1.5244 ∈ 10^{-4}/ 0.494 = 3.0584 × 10^{-4}

From Moody's diagram, friction factor (f) = 0.016

Frictional head loss, h_f = flsv²/(2gd)

Taking length of suction pipe = 6 m and delivery pipe = 14 m, static suction and delivery heads 6 m each, h_{fs} = 0.016×6×22/(2×9.81×0.494) = 0.0396 m, foot valve loss from fig-1 = 0.6 m

Total suction head = 6 + 0.04 + 0.6 = 6.64 m

Delivery side

R_e = V D/ν = 3 × 0.4033/1.054 × 10^{-6} = 1.148 × 10^6

∈/ D = 1.5244 × 10^{-4} / 0.4033 = 3.7798 × 10^{-4}

Friction factor, f from Moody's diagram = 0.0165

Equivalent lengths:

The equivalent lengths are obtained from table.

Gate valve = 9 m, 900 long radius elbow = 27 m

Therefore, effective delivery pipe length = 14 + 9 + 27 = 50 m

Head loss on the delivery side, h_{fd} = (0.0165 × 50 × 32) /(2 × 9.81 × 0.4033) = 0.9384 m Velocity head = 32 / 2 × 9.81 = 0.4587 m

Total delivery head = 6 + 0.9384 + 0.4587 = 7.397 m

Total head = 6.64 + 7.4 = 14.04 m

Required power = γQ/75 η=(1000×0.3833×14.04)/(75×0.4) =183.661 HP

References

Lal, J. 1973. Hydraulic Machines, Metropolitan Book Co. Pvt. Ltd. Delhi.

Mal, B.C. 1995. Introduction to Soil and Water Conservation Engineering. Kalyani Publisher. New Delhi.

Michael, A.M. and Khepar, S.D.1990. Irrigation Wells and Pumps. Tata McGraw Hill., New Delhi.

Ram, S., Sharma, H.C. and Chauhan, H.S. 1982. Pantnagar Propeller Pump. G.B.P.U.A.T., Pantnagar.

2

Aerators for Fish Culture

Aeration is the addition of oxygen or air containing oxygen to water. Any aquatic species important for commercial purpose requires oxygen. Even photosynthetic plants need oxygen when it is dark.

2.1 Need of aeration

1. Rise in atmospheric temperature causes increased rate of biological degradation of organic matters and subsequent depletion in dissolved O_2 concentration in water.
2. Prolonged cloudy day may stop photosynthetic oxygenation and subsequent depletion in dissolved oxygen (DO).
3. In case of intensive and semi-intensive type of culture system, a greater amount of oxygen is required for the respiration of all aquatic species causing depletion of DO.
4. For the above type of culture, during water exchange it is required to mix the incoming water. For this purpose, feed distribution and waste disposal, circulation of water is essential. Aeration helps in mixing and circulation of water.

2.2 Functions of Aerators

It utilizes energy input to:

i) Increase the liquid surface area available for O_2 transfer end/ or to

ii) Mix the liquid with O_2 to ensure that water of low D.O. is brought into contact with O_2 or air.

2.3 Types of Aerators

The aerators used for aquaculture are

1) Gravity aerator,
2) Surface aerator
3) Diffuser aerator and
4) Turbine aerator.

1) **Gravity aerator:** Gravity aerator utilizes potential energy of water to increase gas liquid interfacial area thereby increasing the oxygen concentration of water. They can be natural and manmade.
2) **Surface aerator:** They are used to break up or agitate the surface of water so that oxygen transfer takes place. Examples are i) pump spraying water into air, ii) water pumped through nozzles and iii) hydraulic jump. Commercial surface aerators are i) spray type surface aerator and ii) paddle wheel aerator.

2.4 Design of surface Aerator system

Surface aerators are available in different designs. Much of the designed based on trial and error because only limited amounts of actual design information are available.

Design steps

1) Determine oxygen demands in the system, both chemical and biological demand. Biological demand is the sum of the demand of cultured organisms, the demand of bacteria in any biological filters and the demand of free-floating bacteria.
2) COD: The quantity of O_2 used in biological and non-biological oxidation of materials in water.
3) Determine K_La, the overall oxygen transfer coefficient. The value of K_La (available from the aerator manufacturer) must be adjusted for actual water temperature by the equation

   $$(K_La)_T = (K_La)_{20} \times (1.024)\, T^{-20} \quad (2.1)$$
4) Evaluate α by determining K_La for pure water and for the waste water under similar conditions.

   $$\alpha = (K_La)\text{wastewater} / (K_La)\text{clean water} \quad (2.2)$$

   Similarly,

   $$\beta = \frac{\textit{oxygen saturation concentration of waste water}}{\textit{oxygen saturation concantration of clean water}} \quad (2.3)$$

   Both α and β are best determined by experiment for any waste water. Domestic sewage design calculations use a α value between 0.08 to 0.95 and β value between 0.95 and 1.00.
5) Determine the oxygen concentration gradient during actual aerator operating conditions.

6) Compute oxygen transferred per hour from:

$$OT = K_L a(C_s - C)V \times 10^{-6} \quad (2.4)$$

Where,

OT = oxygen transferred per hour (kg O_2/h),

$K_L a$ = overall oxygen transfer coefficient (h^{-1}),

Cs = actual waste water saturation at operating condition (mg/L),

C = oxygen concentration during operation (mg/L) and

V = volume of water to be aerated (L).

7) Divide oxygen demand by the oxygen transferred per kilo-watt-hour for the aerator selected, to determine the total power requirements necessary to supply oxygen.
8) Divide total power requirement by power rating of individual aerator to determine the number of aerators needed.
9) Locate aerators in the tank such that each aerator has an equal tank surface area in which to operate.

2.4.1 Diffuser Aerators

Diffuses aerators inject air or oxygen into a body of water in the form of bubbles. Oxygen is transferred from the bubbles to the water by diffusion across the liquids film. As the bubbles rise in the water column, there is a relative motion between water and bubbles. This causes water circulation and a renewing of the surface area in contact with the bubble, which increases oxygen transfer. Relatively uniform bubbles having a large transfer surface area per unit gas input increase the overall oxygen transfer.

2.4.2 Turbine Aerator

Turbine aerators consist of a propeller submerged in the water to be aerated. As the propeller rotates, it circulates the liquid, causing greater aeration to occur at the surface. Limited use is made of the turbine aerators as the propeller can easily cause physical damage to the fish. They can however, be used in culture tanks containing non-mobile organisms such as oysters.

2.5 Oxygen Transfer Process

Transfer of oxygen into water can be versed as a three-step process.

1. Transfer of oxygen in the gas-to-gas liquid interface.
2. Transfer across the gas liquid interface.
3. Transfer of oxygen away from the interface into the liquid.

I. The first step is accomplished by a combination of diffusion and convective currents in the gas and is a relatively rapid process.

II. The second step consists of diffusion only and is a relatively slow process. It is the rate which limits the transfer process.

The rate at which a slightly soluble gas such as oxygen is transferred into a liquid is proportional to the area of gas liquid interface and the gradient between the existing and saturation concentration of gas and liquid, the magnitude of liquid film coefficient and the turbulence. Mathematically, these parameters can be expressed as:

$$\frac{dm}{dt} = K_L A(C^* - C) \tag{2.5}$$

where $\frac{dm}{dt}$ = mass rate of change of oxygen with respect to time (kgO_2 per hour),

K_L = oxygentransfer coefficient (cm/h),

a = area of the gas liquid interface, cm^2,

C* = saturation concentration of gas in liquid (mg/L) and

C = existing concentration of gas in liquid (mg/L).

Dividing both sides by volume of water V, eqn.(5) can be written as:

$$\frac{dc}{dt} = k_L \frac{A}{V}(C^* - C) = K_{La}(C^* - C) \tag{2.6}$$

where $\frac{dc}{dt}$ = rate of change of oxygen concentration with respect to time (mg/L-h)

$$= (\frac{dm}{dt} \cdot \frac{1}{V})$$

Now integrating equation (6) for C between C_i and C_t and t = 0 to t, we get

$$\int_{c_i}^{c_t} \frac{dC}{C^* - C} = K_L a \int_0^t dt \tag{2.7}$$

$$\ln \frac{C^* - C^i}{C^* - C^t} = K_L a \times t \tag{2.8}$$

The term $\ln \frac{C^* - C^i}{C^* - C^t}$ in is called log deficit of oxygen transfer.

where $K_L a$ = overall oxygen transfer coefficient (h -1),

Ci = initial concentration of oxygen in liquid (mg/L),

Ct = concentration of oxygen in outlet liquid at any time t(mg/L),

(C*-Ci) = initial oxygen deficit (mg/L) and

(C*-Ct) = oxygen deficit at time t (mg/L).

2.6 Standard Oxygen Transfer Rate (SOTR) Equation

Oxygentransfer rate per unit volume of water at temperature T can be written as:

OTR = $(K_La)_T$ $(C^* - C)$.Standard condition refers to a temperature of 20°C, standard atmospheric pressure, C* = 9.07 mg/L, and C_t = 0 mg/L.

SOTR per unit volume = $(K_La)_{20}$ (9.07 - 0) (2.9)

Standard O_2 transfer rate of the tank water can be computed as:

SOTR = $(K_La)_{20}$ (9.07 - 0).V.10^{-3} in kg O_2/h. (2.10)

V = aeration water volume, m^3.

2.7 Standard Aeration Efficiency

SAE = SOTR/P (2.11)

where P = power consumed, kW

SAE = standard aeration efficiency, kg O_2/kW- h.

If the torque of the aerator shaft T (kg-m) is known then

$$p = \frac{2\pi NT}{6032} \quad (2.12)$$

N = rotational speed of aerator shaft, rpm

2.8 Oxygen Transfer Rate Equation

For standard test of oxygen transfer rate, a temperature of 20°C and initial DO concentration

C_i = 0 mg/L are considered.

Cs* = saturation concentration under standard condition at 20°C for clean water = 9.07 mg/L.

OT_s = oxygen transfer rate under standard condition at 20°C for clean water and initial DO concentration C_i = 0 mg/L (kg O_2/h).

OT_p = oxygen transfer rate of pond water at any temperature T°C and initial DO concentration C_p (Kg O_2/h).

C^* = saturation concentration of clean water at T°C.

Cp^* = saturation concentration of pond water at

$$T°C = \beta C^* \quad (2.13)$$

$(K_La)_s$ = oxygen transfer coefficient for clean water at

$$20°C = (K_La)_{Ts}\,(1.024)^{20-T} \quad (2.14)$$

$(K_La)_{Ts}$ = oxygen transfer coefficient for clean water at

$$T°C = (K_La)s\,(1.024)^{T-20} \quad (2.15)$$

$(K_La)_{TP}$ = oxygen transfer coefficient for pond water at

$$T°C = \alpha\,(K_La)_{Ts} \quad (2.16)$$

If V is the volume of water in m^3, then $OT_s = (K_La)_s C_s^* V \times 10^{-3}$ (2.17)

The oxygen transfer rate in pond water $OT_p = (K_La)_{TP}(C_p^*-C_p)V\times 10^{-3}$ (2.18)

or, $OT_p = \alpha\,(K_La)_{Ts}\,(\beta C^*- C_p)\,V\times 10^{-3} = \alpha\,(K_La)_s\,(1.024)^{T-20}\,(\beta C^*- C_p)\,V \times 10^{-3}$ (2.19)

From (15), $(K_La)_s V \times 10^{-3} = OT_s /Cs^*$ (2.20)

Substituting (18) in (17), $OT_p = \alpha\ OT_s\,(1.024)^{T-20}\,(\beta C^*- C_p)/C_s^*$ (2.21)

Writing $C_s^* = 9.07$mg/L, $OT_p = \alpha\,OT_s(1.024)^{T-20}(\beta C^*-C_p)/9.07$ (2.22)

Dividing both sides by power input, $N = N_0\,[\alpha\,(1.024)^{T-20}\,(\beta C^*- C_p)]/9.07$ (2.23)

where N = oxygen transfer efficiency under pond condition (kg O_2/kW-h) and N_0 = oxygen transfer efficiency under standard condition (kg O_2/ k W- h) = standard aeration efficiency and $OT_p/ OT_s = (N / N_0)$

Writing $\beta C^* = C^*$, $N = N_0\,[\alpha\,(1.024)^{T-20}\,(Cp^*- Cp)]/9.07$ (2.24)

2.9 Aeration in Shrimp ponds

Shrimp ponds with low initial stocks density such as 5 to per m^2 will not require any aerator. Partial water exchange alone will give good result. For ponds with initial stocking density between 10 to 12 per m^2 and targeted production level of 4 to 4.5 t/ha with 2 crops, application of aerators could be avoided by providing sufficient water exchange which may vary from 5% at the beginning to 50% towards harvest. However, in ponds with initial stocking density of 15 per m^2 and above, aerators are essential. In such cases, depending upon stocking density and various other factors ponds are usually provided with 6 to 8 number of aerators for each ha. of water area. Water exchange is normally 20 to 30%. If initial stocking density is very high, say 100/m^2 it may become essential to provide each such ponds of water spread area of say, 500 m2 with at least two aerators. In such cases, requirement of water exchange too becomes very high. Ponds, where feeding rates are above 40 to 50 kg/ha/d, frequent aeration is needed.

2.9.1 Placement of Aerators in ponds

Proper placement of aerators in ponds plays an important role on efficient mixing of water throughout ponds. In a rectangular pond, the best position is to place the aerator at the middle of one of the long sides of the pond directing water parallel to the short side of the pond. Placement of aerator in the corner of the pond to direct water diagonally across the pond is probably the worst position. Depending upon the sizes and shapes of the pond, the placement patterns of the aerators aimed to achieve the best water circulation will vary. Water circulation in the pond is important in breaking of the stratification of water parameters such as O_2, temperature and salinity. Positioning of the aerator on the pond should be decided taking into consideration the prevailing wind direction so as to effectively distribute the oxygen saturated water all over the pond and to promote water circulation in it.

2.9.2 Computation of Aeration Requirement and Number of Aerators

O_2 demand of the water body is broadly classified into 3 categories:

1. O_2 demand of cultured species – fish/ shrimp in pond water.
2. O_2 demand of the planktons.
3. O_2 demand of the benthos (organisms in sediment and mud).

O_2 demand due to shrimp respiration varies with temperature and salinity of pond water. At a temperature of 30°C, the demand can be computed using the equation

$$O_d = 0.487 \times W^{0.881} \qquad (2.25)$$

where O_d = O_2 demand of shrimp (mg/ h) and

W = body wt. of shrimp (g).

The average O_2 demand in pond water for respiration of plankton can be taken as 0.134 mg/L/h for 1 m deep ponds. The O_2 demand for benthos (respiration) can be taken as 0.473 mg/L/h for extensive ponds. For well-prepared pond bottom of intensive ponds this value may be taken as 0.061 mg/ L/h.

Example problems

Compute the requirement of 1 HP paddle wheel aerator for a shrimp farm with following specifications

1. Water spread area – 1 ha
2. Mean water depth – 1m
3. Stocking density – $20/m^2$
4. Survival rate – 70%
5. Avg body weight of individual shrimp – 30gm
6. Avg water temperature – 30°C
7. Saturation concentration of O_2 in the system = 6.94 ppm
8. Minimum dissolved O_2 maintained in the pond = 5 ppm

Solution

Calculate O_2 demand in the water body

Requirement of O_2 for an individual shrimp

$= 0.487 \times W^{0.881}$

$= 0.487 \times (30)^{0.881}$

$= 9.75$mg/hr

Number of shrimps = $20 \times 10000 = 200000$ nos.

O_2 Requirement for 20000 shrimps = IS×SR×SD×WSA

IS = Individual shrimp Requirement = 9.75mg/hr

SR= Survival rate = 70/100 = 0.7

SD = Stocking density = 20/m2

WSA = Water spread area = $1000m^2$

O_2 Requirement for all shrimps = $9.75 \times 0.7 \times 20 \times 10000 = 1365000$ mg/hr

Volume of water in the pond = WSA × depth

$= 10000 \times 1$

$= 10000m^3$

$1m^3 = 1000$ litre

$10000m^3 = 10000000 = 1$ Crore litre

Oxygen Requirement of water body towards animal

$$\text{Respiration} = \frac{1365000}{10000}$$

$= 136.5 mg/m^3/hr$

Animal respiration = 0.1365 mg/l/hr

Animal respiration (AR) = 0.1365 mg/l/hr

Plankton respiration (PR) = 0.134 mg/l/hr

Benthos respiration (BR) = 0.061 mg/l/hr

Oxygen demand = AR+PR+BR = 0.1365+0.134+0.061 = 0.3315 mg/l/hr

Total oxygen demand = 3315000 mg/hr

Conversion of mg/l/hr to kg/hr

$= 0.3315 \times 10000000$

$= 3315000 mg/l/hr$

$= 3315000/1000 \times 1000$

Oxygen demand = 3.315 kg/hr

Oxygen Transfer Rate

$OTR = (OT)_{20} \times [\beta(cs\text{-}c)/8.84] \times (1.024)^{t20} \times \alpha$

$= 1.8 \times [0.98(6.94\text{-}5)/8.84] \times (1.024)^{10} \times 0.90$

$= 1.8 \times 0.2151 \times 1.2677 \times 0.90$

$= 0.4417 kg\ o_2/kw/hr$

No. of aerators Required for the given pond

$$= \frac{OD}{OT} = \frac{03.315\, kg\, o_2 / hr}{00.4417 kg^{\frac{o^2}{kw}/hr}} = 7.5kw$$

1HP = 746w

1HP = 0.746kw

$$= \frac{7.5}{0.746} = 10.05 = 10\ HP$$

Result

10 number of 1 HP motor is required for aeration in the 1 ha water spread area.

(1) Calculate the standard oxygen transfer rate and standard aeration efficiency of an aerator of capacity 2 HP (1 .5-KW). Following Is the data obtained during a standard test conducted in a clean tap water basin of capacity 200 m³. The test was run sufficiently long to

determine that $(Cs)_{25}$ of the basin water was 6.8 mg/lit.

DO at 20% saturation = 1.36 mg/lit in 11.2 min

DO at 80% saturation = 5.44 mg/lit in 53 min

"θ" factor = 1.024,

(Cs)20= 9.08

Solution: First find out overall oxygen gas transfer coef

$$(K_L a)_T = \frac{Log\,OD_1 - LogOD_2}{(t_2 - t_2)/60}$$

$$\therefore (K_L a)_{25} = \frac{Log(6.8-1.36) - Log(6.8-5.44)}{(t_{30} - t_{20})/60}$$

80% and 20% DO saturation was reached in 53 minutes and 11.2 minutes respectively

$$\therefore (K_L a)_{25} = \frac{Log5.44 - Log1.36}{(53-11.2)/60}$$

$$= \frac{Log5.44 - Log1.36}{0.697}$$

This should be corrected to "θ" factor

$(K_L a) = (K_L a)_{20} \times \theta^{T-20}$

$(K_L a) = (K_L a)_{20} \times \theta^{25-20}$

$(K_L a)25 = (K_L a)_{20} \times (1.024\ )^{25-20}$

$1.98 = (K_L a)_{20} \times \theta^{T-20}$

$$(K_L a)_{20} = \frac{1.98}{(1.024)^{25-20}} = 1.76 \text{ Kg } O_2/\text{hr}$$

∴ Standard oxygen transfer rate

SOTR $= (K_L a)_{20} \times (C_S)_{20} \times V \times 10^{-3}$

$= 1.76 \times 9.08 \times 200 \times 10^{-3}$

$= 3.2$ Kg O_2 /h r

Standard Aeration efficiency

$$SAE = \frac{SOTR}{Power\ in\ put}$$

$$= \frac{3.2}{1.5} = 2.13 kgO_2/KW/hr$$

(2) A standard test conducted on a 1.5 kw (2-hp) aerator yielded the data as given below. The test tank contained 5000m³ of clean tap water . The test was run to determine that, the $(Cs)_{25}$ of the basin was 6.8 mg/lit. 20% and 80% saturation were considered.

DO at 20% saturation = 1.36 mg/lit in 11.2 min

DO at 80% saturation = 5.44 mg/lit in 53 min

Find SOTR and SAE

Solution: $(KLa)T = \frac{LogOD_2 - LogOD_2}{(t_2 - t_1)/60}$

$$= \frac{Log(6.8-1.36) - Log(6.8-5.44)}{(53-11.2)/60}$$

$$= \frac{Log\,5.44 - Log1.36}{0.697}$$

= 1.98 Kg/hour

$\therefore (K_L a)_T = (K_L a)_{20} \times \theta^{T-20}$

$$\therefore (K_L a)_{20} = \frac{1.98}{(1.024)^{25-20}} = 1.76/hr$$

$SOTR = (K_L a)_{20} \times (C_S)_{20} \times V \times 10^{-3}$

= 1.76 × 9.08×5000×0.001

= 79.905 kg O_2/kw/hr

(3) Calculate the actual (field) oxygen transfer rate and field aeration efficiency of an aerator with rated (SOTR)20 of 25 Kg O_2/hr. the measured dissolved oxygen concentration of pond water at 25°C and salinity 15ppt is 4.00 mg/lit. The power consumption is 9.00 Kw.

Solution: **Assume *a* = 0.90**

θ = **1.024**

β = **0.98**

Cs at 25°C & ppt 15 = 7.57

$$(OTR)_f = \frac{\infty(SOTR)_{20\times\theta}T(\beta Cs - Cm)}{9.092}$$

$$= \frac{0.9 \times 25 \times 1.024^{25-20} \times (0.98 \times 7.57 - 4)}{9.092}$$

$$= \frac{25.33 \times (3.42)}{9.092}$$

$= \quad 9.53$ Kg O_2/hr

$$(\text{AE})_f = \frac{(OTR)t}{Power} = \frac{9.53}{9.00}$$

= 1.058 Kg O_2/KW-hr

(4) The field oxygen transfer rate of an aerator of power consumption capacity 8.6 Kw is 18 Kg O_2/hr. The measured DO concentration of pond water at 30°C, salinity 15 ppt and pressure (BP) is 4.5 mg/lit. The saturated DO concentration at test temperature is 6.94 mg/lit. Assuming α as .0.90, β as 0.98 and θ = 1.024, calculate the standard oxygen transfer rate (SOTR) & standard aeration efficienty (SAE).

Solution:

$$(\text{OTR})_f = \frac{\infty(SOTR)20 \times \theta^{T-20} \times \beta Cs - Cm}{9.092}$$

$$18 = \frac{0.92(SOTR)20 \times (1.024)^{30-20} \times (0.98 \times 6.94 - 4.5}{9.092}$$

$$18 = \frac{0.92(SoTR)_{20\times1.23} \times 2.30}{9.092}$$

$18 \times 9.092 = (\text{SOTR})_{20} \times 2.55$

$\therefore (\text{SOTR})_{20} = 64.18$ Kg O2/hr

$$\text{SAE} = \frac{64.18}{8.6} = 7.46 kgO_2/KW/hr$$

(5) What is the field oxygen transfer rateand aeration efficiency of a surface aerator if the pond DO is 4.0 mg/lit at temperature 25°C and power consumption is 8.60 Kw . The (SOTR)20 is 19 Kg O_2/hr.

Solution: Please refer to table given below

DO Mg per liter	Water Temperature in Degree Centigrade					
	10	15	20	25	30	35
0	0.71	0.71	0.72	0.74	0.76	0.79
1	0.64	0.64	0.64	0.65	0.66	0.67
2	0.57	0.53	0.55	0.55	0.55	0.55
3	0.51	0.49	0.47	0.45	0.44	0.43
4	0.44	0.41	0.38	0.36	0.33	0.30
5	0.37	0.34	0.30	0.26	0.22	0.18
6	0.31	0.26	0.21	0.17	0.12	0.06
7	0.24	0.19	0.13	0.07	0.01	0.00
8	0.17	0.11	0.04	0.00		
9	0.11	0.04	0.00			
10	0.04	0.00				

$$(OTR)_f = f \times (SOTR)_{20}$$
$$= 0.36 \times 19$$
$$= 6.84 \text{ Kg/hr}$$
$$\therefore (AE)_f = \frac{6.84}{8.6} = 0.8 \text{ Kg } O_2\text{/Kw-hr}$$

(6) Calculate the number of aerators required for one hectare semi-intensive shrimp culture pond as per the data given below

Pond is stocked with : 1,50,000 Nos. of shrimp

average weight of shrimp after 60 days culture @ 70% survivelity :15 gms Salinity of

pond water at a temperature of 30°C : 15 ppt

Saturated oxygen concentration required : 4 mg/lit

Standard oxygen transfer rate of 1 HP paddle wheel aerator : 1.8 KgO_2/kw/hr

Assume plankton and sediment respiration as 0.134 mg/lit/hour and 0.061 mg/lit/hour respectively.

Solution :

Oxygen demand = Planktonre spiration+Sediment respiration+Shrimp respiration

∴ Shrimp respiration (R) = $0.487\,(W)^{0.881}$

= $0.487\,(15)^{0.881}$

= 5.29 mg/hour

∴ Total shrimp respiration = $5.29\times1{,}50{,}000\times0.7$= 5,55,450 mg/h

∴ Total shrimp respiration = 10,000

= 55.55 mg/m^3/hr

= $= \frac{\infty()}{}$ = 0.056 mg/lit/hr

∴ Total oxygen demand = 0.056 + 0.134 + 0.056

= 0.251 mg/lit/ht

= 0.251 gm/m3/hr

∴ For one hectare pond = 0.251 × 10,000

= 2.51 kg/hr

Saturated dissolved oxygen concentration T = 30°C and Salinity 15 ppt

= 6.94 mg/lit

Field oxygen transfer rate = $(OTR)_f$

$$(OTR)_f = \frac{\infty(SOTR)20\times\theta^{T-20}\times\beta Cs-Cm}{9.092}$$

$$= \frac{0.90\times1.8\times(1.024)^{30-20}\times(0.98\times6.94-4)}{9.092}$$

$$\frac{1.62\times(1.024)^{10}\times2.80}{9.092}$$

$$\frac{1.62\times1.27\times2.80}{9.092}$$

= 0.63 kgO_2/kw-h

$$\text{Total power requirement of aerator} = \frac{2.51 KgO_2/hr}{0.63}$$

= 3.98KW

Capacity of one aerator = 1 HP = 0.746 KW

$$\therefore \text{ No. of aerators required } = \frac{3.98}{0.746} = 5.3$$

Say 6 Numbers of 1 HP each aerators are required.

References

Mal, Bimal Chandra. Aquacultural Facilities and Equipment. Academic Press, 2021.

Jobling, M. J. H. Tidwell (ed): Aquaculture production systems. Aquacult Int 21, 733–735 (2013).

3

Feeders for Aquaculture

The feeding is an important operation in aquaculture, it is the process of providing nutritious supplements or live feeds to the culture animal. The equipment used for this purpose is named as feeders. Equipment that is connected with fish feeding can be divided into three major groups, as follows:

- Feed storage facilities
- Feed transporting equipment
- Feed distributors

3.1 Fish Feed Storage Facilities

In fish-feed storage the same facilities can be applied that are widely used in animal husbandry. In large fish farms and in integrated farms there can be a central or primary feed store and individual secondary storage facilities at the fish ponds. The primary store is usually a building in the center of the farm, while the secondary storage facilities are silos with a capacity of 5-20 m^3. These silos can be filled by mechanical bucket elevators or pneumatic. The feed can be discharged into the feeding boats or into other feed distributors through a discharge pipe by gravity or by a discharging screw. The application of this latter type is limited as the discharging screw needs an electric energy supply. The storing of different feed grains in silos is trouble-free and convenient, however the storage of coarse-meal type feed materials tends to make them arch which can make discharge difficult. When a feed storage silo is decided to be installed on a fish farm, adequate foundation should be designed and accessibility ensured.

3.2 Feed Transporting Equipment

Feed transport is a complex operation that forms a link between feed production, preparation, storage and distribution. Those facilities and equipment therefore, that are connected with feed transport have to be designed carefully. Within the fish farm different transporting methods can be taken into account as follows:

- Waterway
- Paved road
- Railway
- Cableway
- Earth road
- Combination of the above

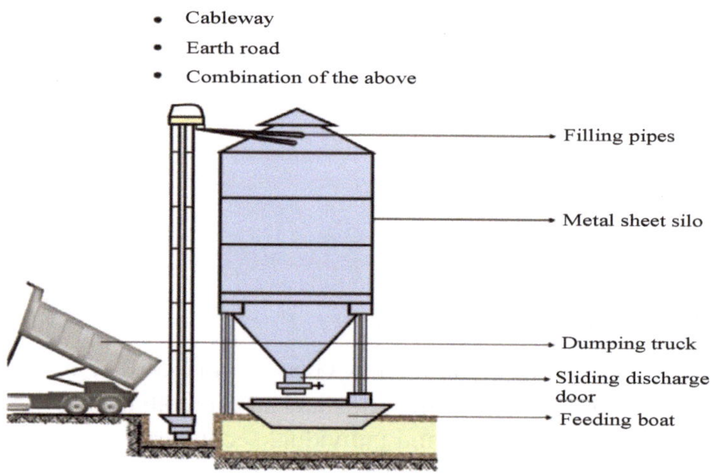

Fig. 3.1: Feed storage silo in fish farm (USSR)

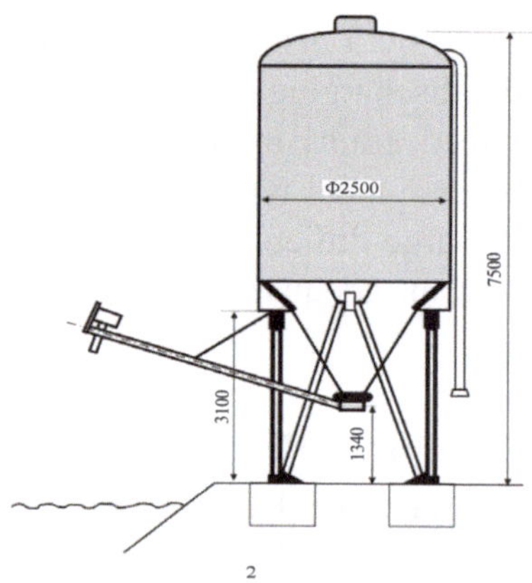

Fig. 3.2: Type Hungaria 20 fibre glass feed storage silo with unloading screw (Hungary)

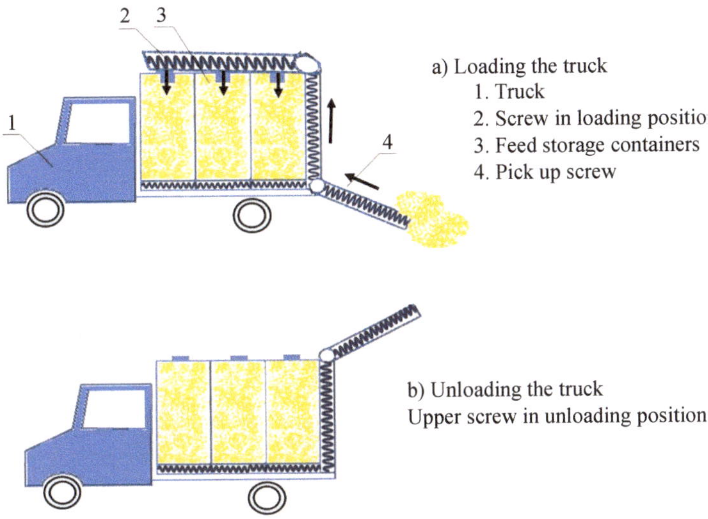

Fig 3.3: Type ESZK 10 feed transporting truck with mechanical loading screw (USSR)

For feed transport, all vehicles, and loading and unloading equipment can be used that are available on agricultural farms. The most common feed transporting vehicles are different types of trucks.

- The trucks have to meet certain requirements, as follows:
- Ability to move on rough and wet surfaces (taking into account fish pond conditions)
- Tipping platform
- Power take-off for driving different adaptors (loading, unloading, etc.)
- Ability to tow a trailer

For feed transport special vehicles have been developed with a large-size container and feed discharging device that can be mechanical (Figure 3.3) or pneumatic.

Feed transporting trailers pulled by multi-purpose tractors are also common on fish farms. A special tractor-pulled feed transporting trailer can be seen in Figure 3.4. For loading the feed transporting vehicles, different loaders can be applied.

1. Air blower
2. Oil separator
3. Check valve
4. Pressure meter
5. Flip valve
6. Feed container

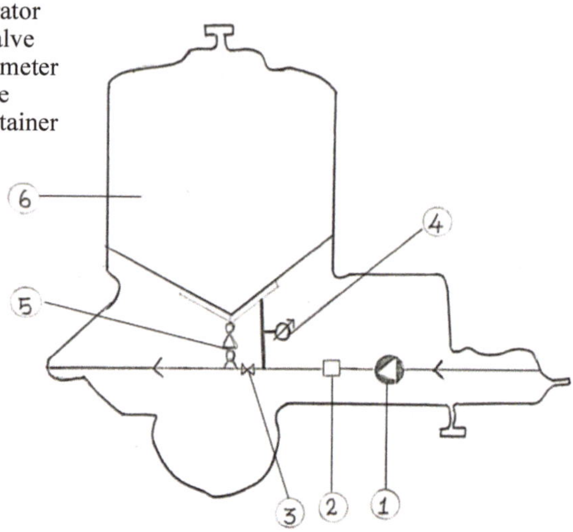

Fig. 3.4: Tractor pulled feed transporting and filling cart

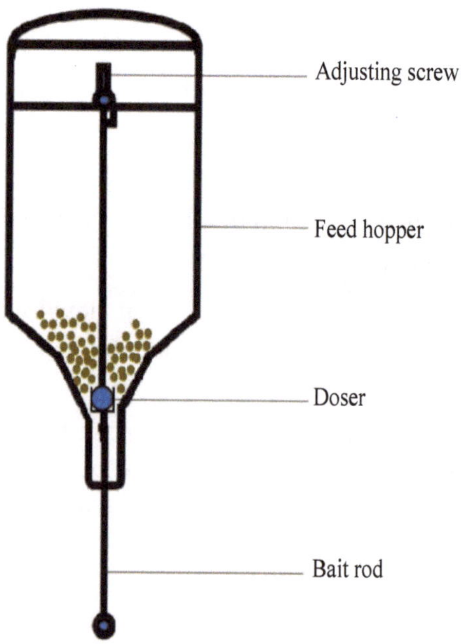

Fig. 3.5: Pendulum type demand feeders (a) MORITZ demand feeder (GDR)

3.3 Feed Distributors

Feed distributors are specially designed feeding equipment. Whilst for feed storing and transport, normal agricultural equipment can be used, for fish-feed distribution, specially designed feeding equipment is needed. There are two major ways in which fish feeding technology, and thus equipment, differ from those generally used in animal husbandry.

- the feed has to be placed in water
- very small-size feed is used.

At present there are many different types of feeding equipment being used world-wide. The feeders can be divided into groups according to the fish species farmed, e.g. special feeders have been developed for eel; age of the fish (different types of feeders are needed for fry, fingerlings and market fish); type of feed (dry and wet feed dispensers). However, these types are based on a few basic principles. Fish feeders can be divided into two major groups, as follows:

- Stationary feeding equipment
  1. without supplied energy; demand feeders
  2. with supplied energy; automatic feeders (electric, pneumatic, hydraulic)
- Mobile feeding equipment
  1. feeding carts
  2. feeding boats

3.3.1 Stationary Feeding Equipment

Demand feeders

Demand feeders are controlled by the fish themselves according to their appetite. These feeders are easy to install and operate. The level of the feed in the container is a good indication of the condition of the fish. Demand feeders can be divided into two main groups, namely bait-rod or pendulum-type feeders and submerged plate-type feeders. Different pendulum-type feeders are shown in **Figure 3.5,3.6,3.7.** These feeders are not suitable for young fish since they are not able to operate them. The outlet size of these feeders should be adjusted carefully, otherwise uncontrolled pouring of feed can be caused by the spontaneous touching of the fish.

The submerged plate-type demand feeders are controlled not by the contact of the fish, but the removal of all feed from the plate **(Figure 3.8).** This type of feeder can be operated by small fish as well; however, their operation is more troublesome, and are not as popular as the pendulum-type feeders.

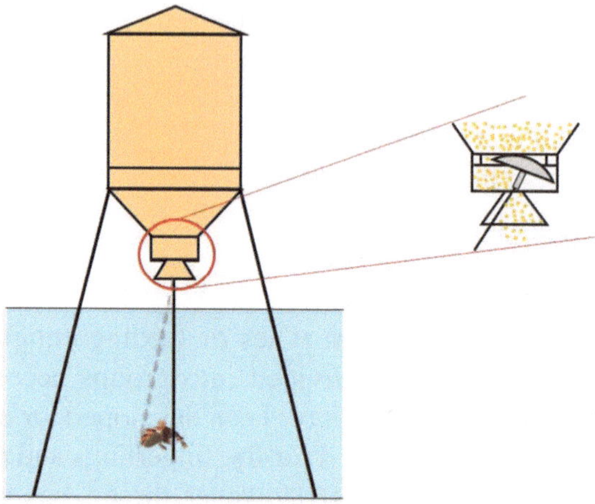

Fig. 3.6: Pendulum type demand feeders (c) EWOS demand feeder (Sweden)

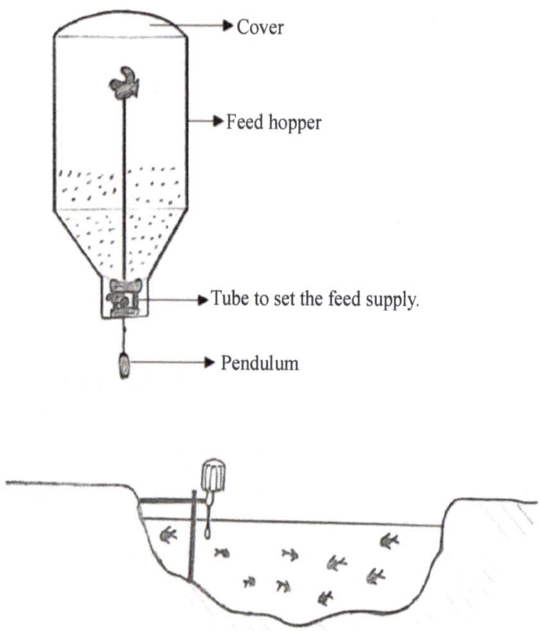

Fig. 3.7: Pendulum type demand feeders (b) UKA-25 demand feeder (USSR)

Automatic feeders

Automatic feeders can be classified according to their energy supply as being electric, pneumatic and hydraulic-type feeders.

(a) *Electrically operated automatic feeders*

- These feeders consist of three main parts; feed container; feed distributing and portioning device; timber.
- The designs differ from each other in the applied feed distributing device, and can be grouped as follows:
  a. electromagnet operated with alternative movement
  b. electromotor operated with revolving movement
  c. vibrator powered

Two electromagnet-operated feeders are shown in Figure 3.9. When the timber operates the electromagnet, it slides apart and opens an outlet under the container through which feed falls into the fish tank. Various designs of electromotor-operated automatic feeders are there. In some electromotor-operated automatic feeders, the revolving plate placed under the container moves slowly and the feed falls directly down into the water. In some cases, instead of a vertical shaft revolving plate, a horizontal shaft screw and belt is used for discharging and portioning the feed into the fish tank (Figure 3.10). When the revolving plate rotates quickly, the feed is spread over a large surface of the water. Recently vibrator powered feed distributors have been developed. In another type of vibrator, when the vibration is generated by a small electric motor with an unbalanced rotor. The electric automatic feeders are controlled by a timer with which the length of the feeding time and the time interval between two feedings can be adjusted. A single control unit can be used for one individual feeder or a central control unit can operate more feeders.

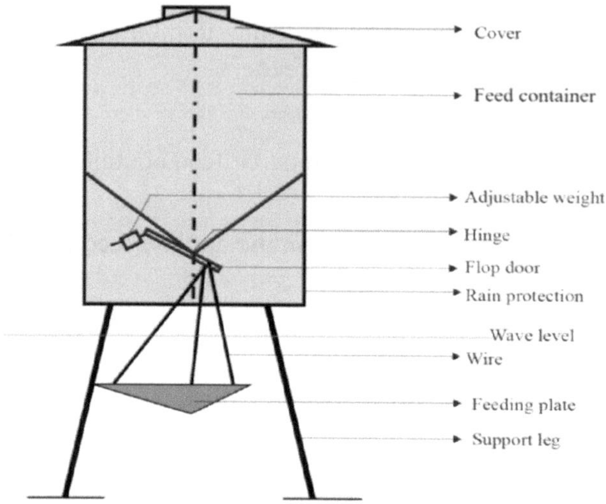

Fig. 3.8: Submerged plate type demand feeder

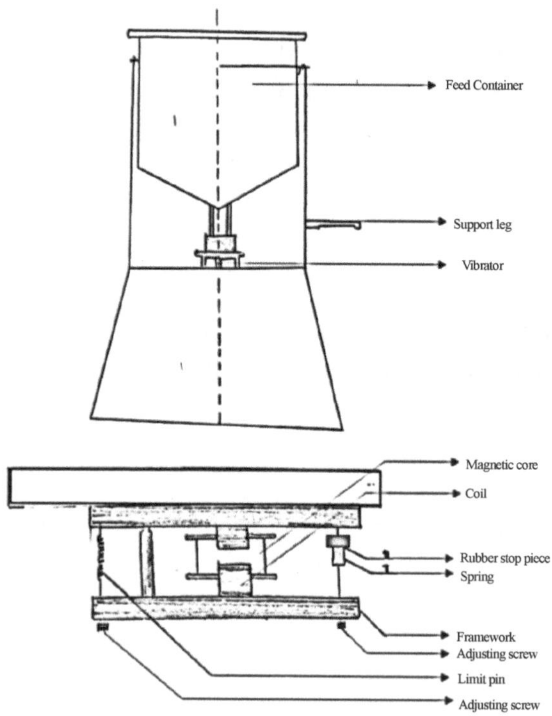

Fig. 3.9: Automatic feeders with vibrator powered distributing device invertical direction (Hungary)

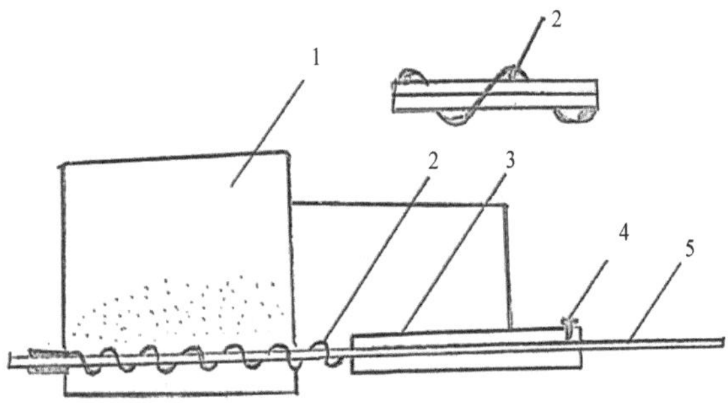

Fig. 3.10: Electromotor operated automatic feeders - Screw type fish feeder (FRG)

(b) *Pneumatic-type automatic feeders*

These feeders shoot the feed by means of compressed air, spreading it over the water surface. The typical designsare shown in Figures 3.11 a and b. The compressed air supply of the feeders is usually ensured by a central air supplying system, but the feeders can be controlled by a separate electric system centrally, or individually as well.

(c) *Hydraulic-type automatic feeders*

These feeders can be used on farms where electric energy supply is not available, but where there is a continuous water flow. As shown in Figure 3.12, the discharging screw is driven by a water wheel. The rate of feed portioning can be adjusted by changing the gear drive. Some hydraulic feeding systems are known where the water is used not only for driving a mechanical feed distributing device, but for transporting the feed into the fish culturing unit through a pipeline. Hydraulic feeding system is used for feeding nursery ponds. These systems can also be applied for distributing wet food like chopped trash fish, etc.

3.3.2 Mobile Feeding Equipment

- Feeding carts
- Feeding boats

This group of feeders can also be divided into two major groups: feeders distributing the feed along the side of the pond and feeding equipment applying feed from a boat.

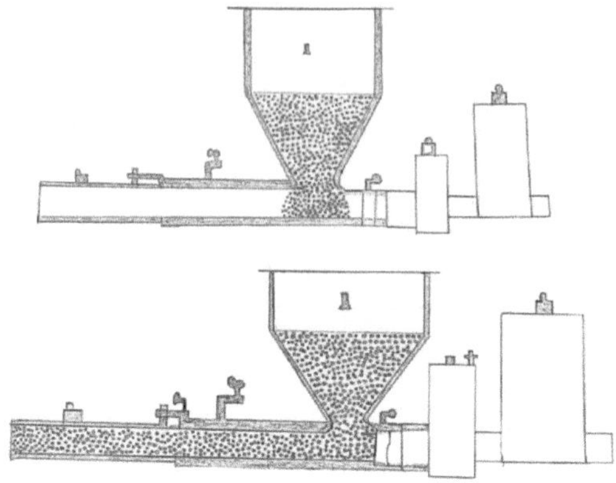

Fig. 3.11(a): EWOS pneumatic type automatic feeder (Sweden) a) The scheme of the feeder

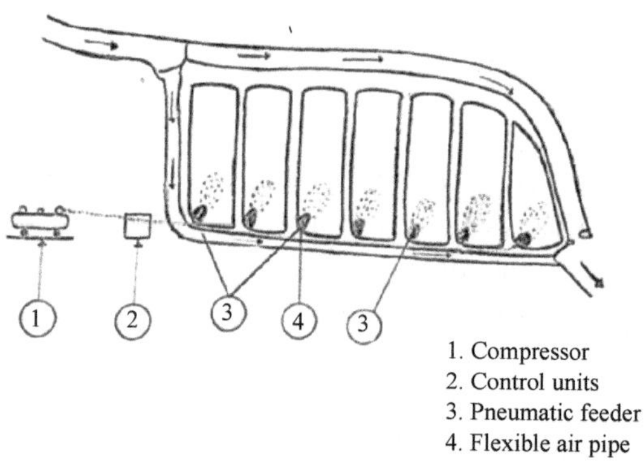

Fig. 3.11(b): EWOS pneumatic type automatic feeder (Sweden) b) The scheme of the pneumatic feeding system

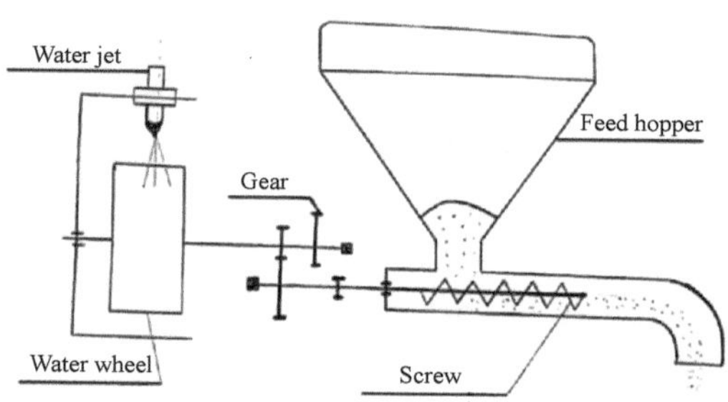

Fig. 3.12: Water-wheel type automatic feeder

Feeding carts

Feeding carts can be used efficiently in intensive fish farming with centralized pond systems connected via suitable roads. Their operation is not automatic (unlike the stationary feeders) but fully controlled by the operator. The feeding carts can be self-propelled or tractor pulled. The basic working principle of these feeding carts is the same; the feed is blown into the fish pond by a turbo blower or large-capacity ventilator. A tractor-pulled feeding cart is operated by an individual gasoline engine and can be controlled from the tractor by an electric remote-control panel. In a self-propelled feeding cart where the feed distribution and portioning can be controlled by a mechanical transmission.

Feeding boats

In fish farms having large ponds, feeding boats can be used efficiently. There is a longitudinal slot in the bottom of these boats through which the feed grain is washed out by water current during the movement of the boat **(Figure 3.13)**. These boats can also be used for fertilization of fish ponds. They are, however, not suitable for applying pelleted feed because the feed swells up in the bottom of the boat and cannot flow out through the slot. In order to make these boats suitable for feeding pellet a pellet-distributing adapter can be mounted on top of the conventional boat.

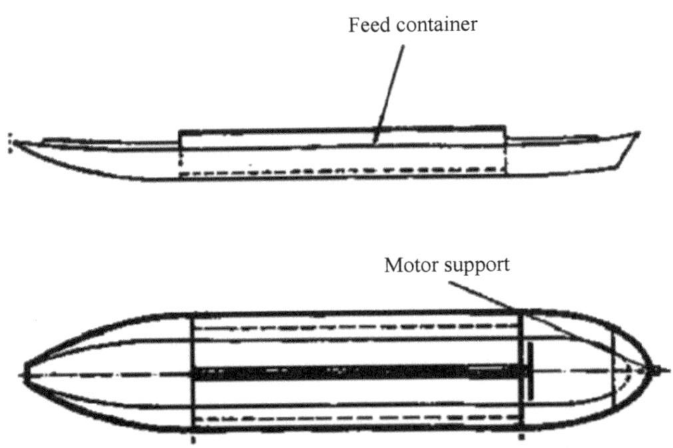

Fig. 3.13: "Lachs" feeding boat with 1 t capacity (GDR)

There are feeding boats that are specially designed for fish farms both for pellet and feed grain distribution. The following figure shows one of these types of feeding boats where the feed is discharged from the containers by screws driven by a paddle wheel that rotates during the movement of the boat.

References

Berka, R., 1973, A review of feeding equipment in fish culture. EIFAC Occas. Pap., (2):32 p.

Kuli, B. and I. Csávás, 1976, A tógazdaságihaltermelésgépesitése (Mechanization of pond fish production). Budapest, AGROINFORM

Tjuktjaeva, J.S. and V. Ja. Lysenko, 1974, Handbook on mechanization of processes of pond fish culture. Moskva, PischevajaPromyshlennost

Torban, S.S., 1977, Mechanization of processes of industrial fisheries. Moskva, PischevajaPromyshlennost

4

Filtration System

4.1 Introduction

Filtration may be defined as the separation of a mixture or solution into component parts. The resulting components may be pure or may be solutions or mixtures. In water/wastewater treatment filtration technique is used for removing suspended solids, dissolved solids and organic matter.

4.2 Types of Filtrations

There are four types of filtration technique which are generally being used in water / wastewater treatment – a) Mechanical, b) gravitational c) chemical and d) biological. In the following section, they are described in brief.

a) Mechanical

In aquaculture, mechanical filtration is used primarily for the separation of solids and liquids. Mechanical filters utilize the differences in particle size of the solution (or mixture) components to extract one part(s) from the other(s). They usually are simple in operation and relatively easy to maintain. If correctly used. Their operating and / or cleaning cost becomes excessive if the concentration of suspended material exceeds reasonable values.

i) Stationary screen

It consists of a stationary screen placed across the fluid flow path such that the fluid must pass through it. Particles larger than the screen mesh cannot pass through the screen and are collected on it.

ii) Rotary screen

Rotary screens are used where screen plugging is a problem. Rotary screens, as the name indicates, are rotating screens constructed such that the screen operates only partially submerged in the water that is to be filtered. There are two basic types of rotary screen: axial flow and radial flow. Fluid passes through an axial flow screen parallel to the axis of rotation of the screen. In a radial flow screen flow travels radially toward or away from the axis of rotation.

iii) Vibratory Screens

Vibratory screens are also used for removal of particles from water. Two types are commonly used: radial flow and axial flow. In the basic design of an axial flow vibratory screen, influent enters one end of the screen and the particulate materials are moved along the screen length by the vibratory motion of the screen. Water flows through the screen and is discharged at the outlet. The screen is mounted on springs and is given vibratory motion by an electromagnetic vibrator, an eccentric drive, or other suitable means. Separation by a vibratory screen is dependent on hole size, feed rate, vibration frequency and amplitude, and properties of the particulate material.

b) Gravitational

Gravitational separation utilizes the force of gravity to extract particles from a fluid. Density differences between the particles and fluid cause the particles to travel downward in a quiescent fluid column. A simple example of gravitational filtration is sedimentation. Sedimentation is the process of allowing particulate materials having density greater than that of the suspending liquid to settle out under gravitational forces in a quiescent or slowly moving liquid. Aggregation of finely divided particles by addition of chemicals and/or other materials to increase settling velocity is referred to as coagulation. Chemical Precipitation, a third related process, may be defined as the addition of a chemical to wastewater, resulting in a chemical reaction that produces an insoluble compound. This insoluble material then settles out by sedimentation. Other gravity filters include centrifuge and hydroclone. Each has possible application in aquacultural enterprises for separating solids from liquids.

c) Biological

Biological filters are devices to culture microorganisms that will perform a given task for us. Different types of organisms will perform different tasks. Part of the art of designing and using biofilters is to create an environment that will promote the growth of the organisms that are needed.

Biofilters are used to help maintain water quality in recirculating or closed loop systems. Biofilters are also used to improve water quality before water is discharged from a facility. There are many different methods of maintaining good water quality and biofiltration is only one component of the total picture. It is however, a very important and essential component especially for recirculating aquaculture or aquarium systems.

Depending on design and application, biofilters have the ability to accomplish the following functions:

1) Remove ammonia
2) Remove nitrites
3) Remove dissolved organic solids

In general, there are three types of aerobic microorganisms that colonize biofilters for aquaculture. Heterotrophic bacteria utilize the dissolved carbonaceous material as their food source. Nitrosomonos sp. bacteria utilize ammonia as a food source and produce nitrite as a waste product. Nitrospira sp. utilize nitrite as a food source and produce nitrates as a waste product. Nitrosomonos and Nitrospira will both grow and colonize the biofilter as long as there is a food source available. Unfortunately, both of these types of bacteria are relatively slow growing. Heterotrophic bacteria grow about 5 times faster and will out compete the other two types for space if food is available. Since most aquaculture biofiltration systems are designed for the purpose of converting and removing ammonia from the water this presents a problem.

There are three ways to deal with this problem. The first is to remove most of the carbonaceous BOD (biological oxygen demand) before the water enters the biofilter. The second method is to provide sufficient extra capacity (surface area) in the biofilter to allow all of the various bacteria to grow. Another method is to have a very long plug flow path through the biofilter. This allows different zones of bacteria to establish themselves in different parts of the biofilter.

There are 4 main types of aerobic biological filters and several subcategories of each. Here is a listing of the major types.

I. Recirculated Suspended Solids (Activated sludge)

II. Aquatic Plant Filters
   a) Unicellular (Microscopic)
   b) Multicellular (Macroscopic)

III. Fluidized Bed Filters
   a) Sand Filters
   b) Bead Filters

IV. Fixed film
   a) Rotating Biological Contactors (RBC)

b) Trickling Filters

V. Submerged Filters (with or without aeration)

a) A.Up flow

b) Down flow

c) Horizontal flow

d) Moving Bed

Anaerobic filters can also be defined as biofilters but they are never the main biofilter used for maintaining water quality in the culture system. It is because they are not capable of effectively cleaning the water to the level required and they operate too slowly. Theoretically, one could use an anaerobic filter to convert the nitrates into nitrogen gas. However, this is a difficult process to control and it is generally less expensive to replace a small amount of water to remove nitrates.

4.3 General Water Quality Maintenance Principles

Not all aquaculture applications have the same requirements for biofiltration. Not only do crops vary in their requirements but different farmers may grow the same crop under different conditions. The biofilter is only one of several components of the system used to maintain water quality. The functions that the biofilter must perform are determined by the presence and effectiveness of other components. Here are some other components and their effects on the system.

4.3.1 Aeration or oxygenating systems

Aeration is always the first step when increasing carrying capacity over an open, lightly loaded system. Mechanical surface aerators, subsurface air bubblers and pure oxygen injection is the typical progression in terms of technology and complexity. All aerobic biofilters require oxygen to operate.

4.3.2 Particulate Filters

Once sufficient oxygen is provided, the next easiest way to improve water quality is to remove suspended solids. This is a more difficult task since particles come in all shapes, sizes and densities. Suspended solids consist primarily of uneaten food and feces which are slightly denser than water. Large particles, above 100 microns, will settle out quite easily. Particles above 50 microns can be filtered out with a screen. Particles below 10 microns are difficult to filter and are generally removed by some other means.

There are many different types of particulate filters that can remove suspended solids. They generally fall into three broad categories.

The first type are settling basins, tube settlers, plate settlers, swirl separators and similar systems that allow the particles to drop out of the flowing stream by gravity. They are relatively simple devices and they work well on large particles. Settling systems generally have very low pump head requirements.

The second type are sand filters, sock filters, drum filters, disk filters, belt filters and similar systems that mechanically remove the particles from a flowing stream. These types of systems "screen" the particles. The size of particle removed is dependent on the size of the screen or sieve. Pump head requirements can vary from low to very high. Some biofilters such as bead filters claim to do both particulate filtration and biological filtration.

The third type of particulate filter is the floatation or bubble seperator. These are commonly known as protein skimmers. In this device, air is bubbled into a column and the fine particles become attached to the surface of bubbles. The resulting froth or foam is collected and removed from the system. These devices require a certain amount of surfactant type compounds in the water in order to work properly.

Removal of suspended solids is important since suspended solids comprise the majority of the BOD (Biological Oxygen Demand). The BOD not removed by the particulate filtration system must be removed by the biofilter before effective ammonia removal will occur. Thus the size of the biofilter is influenced by the effectiveness of the particulate filter.

4.3.3 Foam Fractionators

Foam fractionators are good at removing small particles (under 10 microns) and surface active compounds. They are sometimes referred to as protein skimmers. Since proteins are nitrogenous compounds that degrade into ammonia, foam fractionators can reduce the load on the biofilters. They are definitely useful in systems where water clarity is important. Foam fractionators also add oxygen to the water as a secondary benefit. Unfortunately, foam fractionators do not always work well in fresh water.

4.3.4 Ozone

Ozone is a powerful oxidizer and sterilant. It is potentially harmful to fish, humans and most living organisms. It is definitely harmful to biofilters. It is used to improve water clarity and reduce disease transmission. Ozone should never be used directly before a biofilter. If ozone is used upstream of a biofilter, there should be sufficient retention time after the injection point to ensure that no ozone residual enters the biofilter.

4.3.5 UV light

Certain wavelengths of UV (Ultraviolet) light can be used as a sterilant. UV light is often used with ozone. UV light and ozone are complimentary and synergistic.

4.3.6 Carbon dioxide strippers

Buildup of CO_2 can be a serious problem in a heavily loaded, intensive recirculating system using pure oxygen. The choice of biofilter has a direct influence on the degree to which CO_2 is a problem. In general, any biofilter other than a trickling filter will have a CO_2 problem when pure oxygen is used rather than compressed air for aeration. Building a CO_2 stripper is not a difficult task but it must be included in the overall design of the system. In order to remove carbon dioxide, there must be a large interfacial area between air and water. The interfacial area can be increased through the use of subsurface aeration, mechanical surface aerators, spray systems or packed columns. Subsurface aeration is not very efficient and mechanical surface aerators are difficult to use in an intensive recirculating systems. Spray systems can be big energy users and they are not very efficient either. The best choice for intensive and space limited systems is the packed column. Packed columns can be either cross flow or counter flow systems. Packed columns for CO_2 stripping require fans to either force (push) air in or induce (pull) air through the packing.

4.4 Characteristics of the "Ideal" Biofilter

The following characteristics can be considered a checklist that can be used to rate each of the different types of biofilter.

1) Small footprint - The biofilter should occupy as little space as possible. It is common to have culture tanks and the biofilters under cover for protection and temperature control. Space allocated for biofilters takes away area that could be used for culture tanks.
2) Inert materials of construction - All materials used in the biofilters should be non-corrodible, UV resistant, resistant to rot or decay and generally impervious to chemical attack. In general, marine grade construction materials are required for reasonable working lifetimes.
3) Low capital cost - The biofilter must be inexpensive to purchase or build and cheap to transport to the farm location.

4) Good mechanical strength - The biofilter and its components must be tough enough to withstand the normal wear and tear of a industrial/agricultural environment.
5) Low energy consumption - The energy cost (usually electricity) to operate the biofilters should be as low as possible. The largest energy users are the pumps to move water and compressors to move air.
6) Low maintenance requirements - The biofilters should be self cleaning with little or no care required for the normal life of the crop.
7) Portability - The biofilters should be easily movable to facilitate changes in operation of the facility.
8) Reliability - Ideally the biofilters should have no moving parts that could fail at an inopportune time. If the biofilters does have moving parts, they should be rugged and designed for a continuous operating life of several years.
9) Monitorabilty - It should be easy to observe the operation of the biofilter to insure that it is operating correctly.
10) Controllability - It should be easy to change operating variables to assure optimum performance.
11) Turndown ratio - The biofilters should be able to work under a wide range of water flow rates and nutrient loading levels.
12) Safety - The biofilters should not have any inherent dangers to either the crop or the owner/operator.
13) Utility - The biofilters should accomplish all of the goals set forth in beginning of this paper i.e. removal of ammonia, carbon dioxide, BOD, suspended solids etc.
14) Scalable - A small system should work the same way as a large system. The performance per unit volume should be constant regardless of the size of the system.

4.5 Characteristics of Real Biofilters

4.5.1 Activated Sludge Systems

Activated sludge systems are not common in aquaculture systems. Activated sludge systems are good at removing carbonaceous BOD in systems with high nutrient loadings. They are commonly used in domestic waste water treatment systems. Activated sludge systems are typically expensive to operate and do not provide the effluent water quality necessary for aquaculture.

4.5.2 Aquatic Plant Systems

Plants are not normally used for the primary biofilter in aquaculture systems. They do however provide a very good sink for the nitrates produced by a well-functioning biofiltration system. The marriage of recirculating aquaculture systems and hydroponics are a good example of efficient use of resources. In addition to commercially valuable plants grown in hydroponics systems, aquatic plants such as hydrilla, cattails, water hyacinths and duck weed can be used to absorb nitrates and phosphorus from waste water.

Unicellular plants (algae, diatoms *etc.*) are sometimes allowed to grow in the culture tanks. Some species such as tilapia are tolerant of poor water quality and can use the algae as food. Systems operated this way are sometimes called "green water" systems to distinguish them from the clear water systems that many species require. Green water systems can be a very cost-effective way to culture certain species but they are not recommended for beginners to aquaculture.

4.5.3 Fluidized bed sand filters

Regular sand filters such as the type used for swimming pool filters or potable water filters are virtually worthless as biofilters for aquaculture. The biofilm quickly fills the spaces between the grains and the pressure drop across the filter rises rapidly. Frequent back flushing is required and the active biological film is removed each time. In contrast, fluidized bed filters have been successfully used for aquaculture applications. A sand filter becomes fluidized when the velocity of the water flowing up through the bed is sufficient to raise the grains of sand up and separate each grain from its neighbors. In hydraulic terms, the drag on each particle is sufficient to overcome the weight of the particle and the particle is suspended in the stream of water. The velocity required to fluidize the particle is a function of the shape, size and density of the particle.

Fluidized bed sand filters have several very good advantages. They pack more biologically active surface area into a given volume than any other type of biofilter. In addition, the best shape for a fluidized bed sand filters is a tall column. Thus, they have a small foot print for a given capacity. They are self-cleaning and relatively tolerant of different nutrient loadings.

There are also several disadvantages and potential problem areas with fluidized bed sand filters. The fluidized bed sand filter has a relatively high energy requirement because of the high pressure drop necessary to fluidize the sand. The other main problem with sand filters is that the pressure required to fluidize the bed varies depending on the amount of biofilm on the sand particle. As the biofilm builds on the sand particle the size of the particle increases while the density of the particle decreases. This means that the depth of the bed will tend to increase as the bed ages. It also means that the bed depth will fluctuate as the loading on the bed varies. In order to prevent blowing the sand out of the tank, the tank must be oversized or the flow of water needs to be regulated.

Another potential problem is the uniformity of the water flow. In order to completely fluidize the bed, the water needs to be evenly distributed across the whole bed. Two things can happen if the flow is not uniform. One possibility is that the water will channel and short circuit though the bed. This means that the treatment capacity will plummet. Another possibility is that the short circuit will happen near the wall of the vessel and the abrasive sand will eat a hole through the wall of the vessel.

Fluidized bed sand filters are limited to the oxygen carried in with the water. This means that the water entering the filter should have a high level of oxygen in order to insure a good level of treatment.

4.5.4 Bead filters

Bead filters are a relatively new type of biofilter. They consist of a closed vessel partially filled with small beads of plastic. Usually the vessel is filled with water and the beads float at the top of the vessel. Water flows up through the bed of beads. The beads are small enough to trap most large suspended solids. In addition, the surface of the beads supports the growth of a biofilm. The small size of the beads means that they have a relatively large surface area per unit volume. The more sophisticated systems incorporate a mechanical stirring device such as a propeller on a shaft. Periodically the water flow is shut off and the bed of beads stirred to dislodge the suspended solids. The solids are allowed to settle into the bottom of the vessel and then drained off. This ability to remove suspended solids and act as a biological filter is the main advantage to bead filters.

The difficulty in successfully operating bead filters lies in striking a balance between the competing functions. Too frequent washing to remove solids dislodges the biofilm and disrupts the nitrification process. If the beads are not washed enough however, the solids start to plug the bed. The other potential problem is the presence of large amounts of carbonaceous solids which tends to encourage the growth of heterotrophs at the expense of nitrobacter sp. and nitrospira sp.

Another drawback to bead filters is their relatively high energy consumption due to their high pressure drop. Also, the water flow and pressure drop are not constant. As the bed of beads becomes loaded with solids, the pressure drop rises and the water flow decreases. This leads to cyclic rather than constant performance. Schematic diagram of a bead filter is presented in **Fig. 4.1.**

Since bead filters are not aerated, they are limited to the oxygen carried in with the water. In general this is not a problem since retention times are low. Bead filter systems are probably suitable for small, lightly loaded systems where labor costs are low.

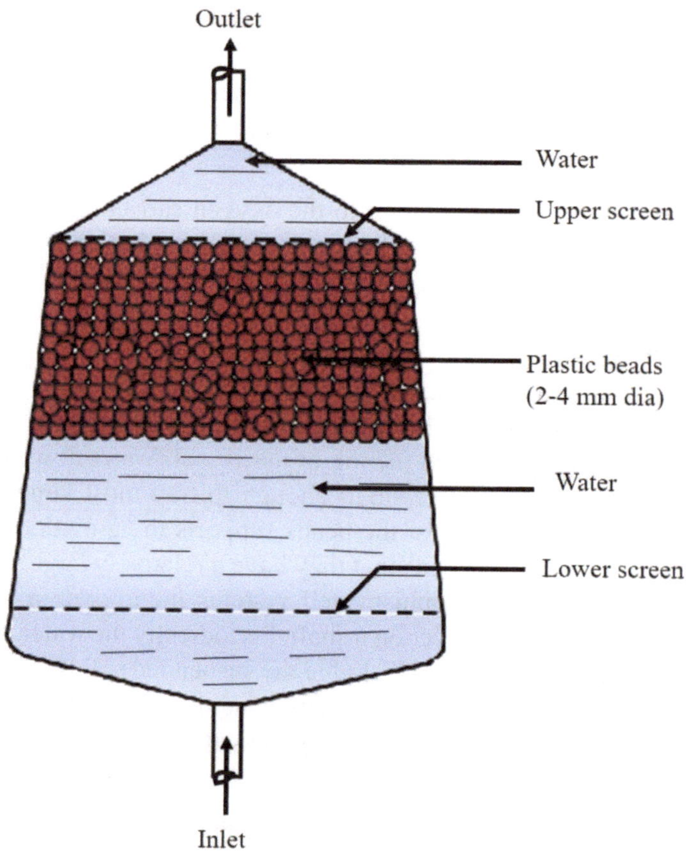

Fig. 4.1: Schematic diagram of a bead filter

4.5.5 RBC (Rotating Biological Contactors)

Like much of the equipment used in aquaculture, RBC's were first used in domestic sewage treatment applications. There are several different types that are manufactured. A typical design consists of plates or disks that are attached to a horizontal shaft. The shaft is located at the surface of the water and it is turned at a very slow speed (1-5 rpm). The disks are half submerged in the water at all times. As they rotate, the biofilm attached to the surface of the disk is alternately exposed to air and then submerged in the water. The original designs used an electric motor to turn the shaft. There is a new design specifically for aquaculture that uses compressed air or pumped water to drive a paddle wheel in the center of the cylinder. These RBC's float in the water and do not require bearings or elaborate mechanical supports.

RBC's have many advantages. They offer excellent treatment efficiencies. They require very little energy to operate and can be located in the culture tank to save space if necessary.

They do not require additional oxygen and are not limited to oxygen contained in the incoming water. They can remove dissolved BOD or ammonia depending on nutrient levels. They are biologically robust and handle shock loads well. It is easy to observe their operation and visually monitor the biofilm. They only have one major drawback besides cost and that relates to reliability. If there is a power failure or the cylinder stops turning for any reason, the biofilm exposed to the air can dry out. When this happens, the cylinder will be unbalanced and become difficult to turn. A typical sketch of an RBC is shown in **Fig. 4.2**.

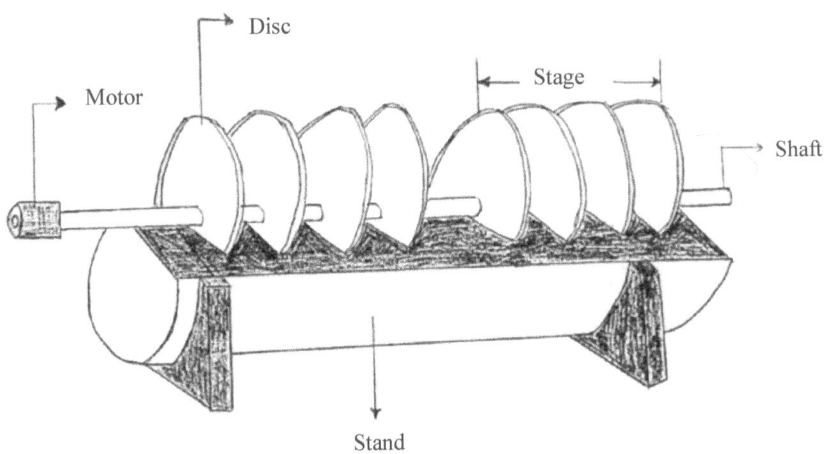

Fig. 4.2. Typical sketch of an RBC unit

4.5.6 Trickling Filters

Trickling filters are one of the oldest types of biological filters. Trickling filter filled with rock or coal was built in the late 1800's for sewage treatment. Trickling filters typically consist of a packing or media contained in a vessel. The water to be treated is sprayed over the top of the media and collected in a sump underneath the media. The surface of the media or packing provides the substrate for the growth of a biofilm.

Trickling filters are rugged and easy to operate. They have the ability to treat a wide variety of nutrient levels. Properly designed systems can handle solids very well. One of the big advantages of a trickling filter is that the water can leave with more oxygen than it entered. Because trickling filters have a large - air water interface, they also act as strippers to remove CO2, H2S, N2 or other undesirable volatile gases. The only major drawback to trickling filters is the energy cost required to pump the water to the top of the filter. A high narrow filter will save space but take more pumping energy. A wide low filter will use less energy but take up more space.

The first step in the design of a trickling filter is to pick the right packing or media. Over the years many different materials have been used for trickling filters but today, the best packing is structured media. Structured media is composed of sheets of rigid PVC that are corrugated and glued together to form blocks.

The most important requirement in the design of any trickling filter is a good water distribution system at the top. A typical schematic diagram of a trickling filter is shown in Fig. 4.3.

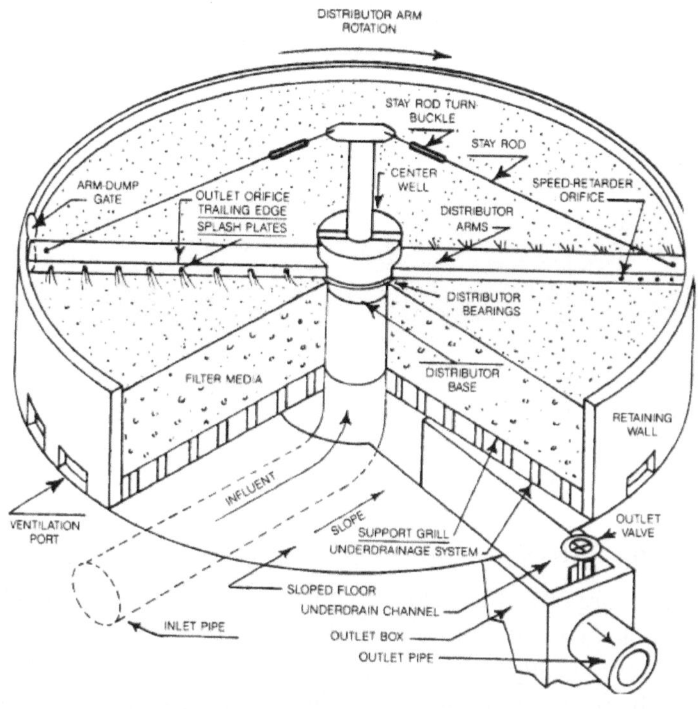

Fig. 4.3. Schematic diagram of a typical trickling filter

4.5.7 Submerged Bed Filters

An under-gravel filter is a classic down flow submerged bed filter. Submerged bed filters have been used extensively for small scale aquaculture and backyard water feature systems. These filters can be operated in up flow, down flow or cross (horizontal) flow. The classic (old) systems consisted of gravel with an under-drain system. An improvement to these systems was the addition of air piping underneath. The air was used to 'bump' the filter to dislodge solids and restore full flow. The problems with these filters are the large size, low void fraction, tendency to plug and extremely high weight. In general, these old gravel-based systems are not suitable for modern aquaculture. Submerged filters can operate with or without aeration. If the flow path is long and the nutrient loading is high, it is wise to have aeration in the filter. One of the easier methods is the traditional aeration system with large silica air stones.

There is always the possibility to install the submerged biofilter media in the culture tank. This has the advantage of saving the cost a separate vessel and associated piping. The big disadvantage to this system is that it is difficult to remove the suspended solids before the water enters the biofilter. Because there are too many different configurations to draw them all, here is a brief description of a few of the possibilities.

1) Air lift the water into one end of a filter designed as a raceway and air lift it back into the culture tank at the other end.
2) Pump the water into a particulate filter such as a rotary drum and then flow through the biofilter.
3) Locate tubes or columns of packing throughout the culture tank and induce a flow through them with air stones.
4) Locate the filter media around the walls of the culture tank and induce a flow up through the media with air stones.

Example Problems

1) **Design low-rate trickling filter for secondary treatment of sewage generated from 10000 persons with rate of water supply 170 LPCD. The BOD_5 after primary treatment is 110 mg/L and BOD5 of final effluent should be 20 mg/L. Consider C = 5.358.**

Solution

$$\frac{S_t}{S_0} = \frac{1}{1+c\left(\frac{D^{0.67}}{QL^{0.5}}\right)}$$

Provide depth D = 1.5 m

Sewage flow = $10000 \times 170 \times 0.80 \times 10^{-3}$ = 1360 m^3/d

Now,

$$\frac{20}{110} = \frac{1}{1 + 5.358\left(\frac{1.5^{0.67}}{QL0.5}\right)}$$

$$0.182\left(1 + \frac{7.03}{QL^{0.5}}\right)$$

$QL^{0.5}$ = 1.562

Q_L = 2.441 m^3/m^2-d where, Q_L = flow/area

Plan area = $\frac{1360}{2441}$ = 557.35 m^2

Hence, Diameter of trickling filter = 26.64 m

2) Design high rate trickling filter for the data given above except effluent BOD_5 = 40 mg/l since polishing treatment is provided after high rate trickling filter. Consider recirculation ratio of 2 and filter depth of 1.8 m.

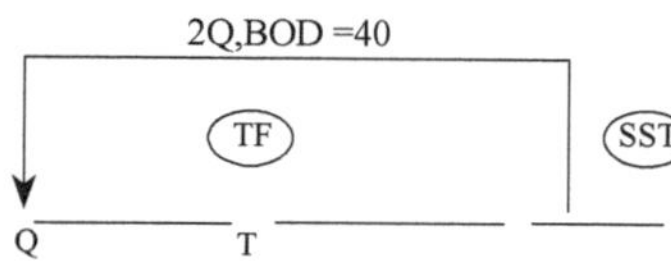

BOD = 110

BOD in the effluent,

$Q \times 110 + 2Q \times 40 = (1 + 2)Q.S_0$

S_0 = 63.33mg/L

$$\frac{40}{63.3} = \frac{1}{1 + 5.358\left(\frac{1.5^{0.67}}{QL^{0.5}}\right)}$$

Q_L = 186 m^3/m^2-d

Waste water flow is 1360 m^3/d

Recycle flow = 2720 m^3/d

and total flow = 4080 m^3/d

Area of Trickling Filter = $\frac{4080}{186}$ = 21.94 m^2

Diameter = 5.285 m

3) **Design a single stage bio tower (super rate trickling filter) for the following data: Average wastewater flow = 500 m³/d,**

Influent BOD = 160 mg/L

BOD removal in primary treatment = 30%, Effluent BOD required = 20 mg/L

Filter depth = 5.0 m; Recycle ratio R/Q = 2

Pilot plant studies using synthetic packing have shown a removal constant K = 2.26 at 20° C, and n = 0.5 (Reynolds & Richard, 1996). The winter wastewater temperature = 15° C. Provide minimum two filters in parallel.

Solution:

$$\frac{S_t}{S_0} = \frac{KD}{e^{QL^{0.5}}}$$

Now, $K_{15} = K_{20}\ 1.035^{(T-20)}$

$= 2.26 \times 1.035^{(-5)}$

$= 1.903$ per day

BOD_5 in incoming wastewater to biotower = 160 (1 - 0.30) = 112 mg/L

Recycle flow = 2Q and BOD = 20 mg/L

Therefore, 112 (Q) + 2Q × 20 = 3Q. S_0

Hence, $S_0 = 50.67$ mg/L

Substituting values in the equation

$$\frac{20}{50.67} = \frac{2.902 \times 5.0}{e^{QL^{0.5}}}$$

Solving we get, $Q_L = 104.8\ m^3/m^2.d$

Wastewater flow = 500 m^3/d

Hydraulic load on filter = 500 + 2 x 500

= 1500 m^3/d

Area required = 1500/104.8 =14.313 m^2

Hence, diameter required for each biotower when two are provided in parallel = 3.02 m

4) A water treatment plant will typically have several filters. Each filter in our calculations will be assumed to have the following specifications.

- Square tank
- Basindepth: 10ft
- Media depth: 2-3ft
- Surface area: <2,100 ft^2
- Filtration rate : 2-10gal/min-ft^2
- Flow through filter : 350-3, 500gpm
- Backwash period : every 24 hours
- Backwash period: 5-10 minutes
- Backwash water: 1-5% of filtered water
- Backwash rate: 8-20gal/min-ft^2
- Filter rise rate: 12-36in/min
- Bed expansion: 50%
- Backwash trough 3ft above media
- Backwash water piped to raw water intake

As you can see, backwashing is a very important part of filter calculations. We will briefly identify some of the backwash characteristics below.

The **backwash frequency** is the same as filter run time. Either term can be used to signify the number of hours between backwashing.

The **backwashperiod** is the length of time which backwashing lasts. The **backwash water** is the water used to backwash the filter. For the filters we're considering, backwash water should be 1-5% of the water filtered during the filter run.

The **backwash rate** is the rate at which water is forced backwards through the filter during backwashing.This rate is homologous to the filtration rate,only with water moving in the other direction through the filter.The backwash rate is typically much greater than the filtration rate.

The **filter rise rate** is the speed at which water rises up through the filter during back washing.This is another way of measuring the back wash rate.

During backwashing, the water pushes the media up until it is suspended inthe water. The height to which the media rises during back washing is known as the **bed expansion**. For example, if the filter media is 2 feet deep, it may rise up to 3 feet deep during backwashing. This is a 50% bed expansion:

$$\text{Bed expansion} = \frac{(3\text{ft} - 2) \quad 100\%}{2\text{ft}}$$

Bed expansion = 50%

Most of these backwash specifications merely describe the type of filter we will be considering and are not used in calculations. However, two factors - the filter rise rate and the backwash period - will be used when calculating the volume of the clear well chamber.

Overview of Calculations

1) Calculate the approximate number of filters required.
2) Calculate the flow through one filter.
3) Calculate the surface area of one filter.
4) Calculate the length of the tank.
5) Calculate the clear well volume

Number of Filters

The treat ment plant's flow should be divided into atleast three filters. You can estimate the number of filters required using the following formula:

$$\text{Number of filters} = 2.7\sqrt{Q}$$

Where:

Q = Flow, MGD

So, for a plant with a flow of 1.5 MGD, then the approximate number of filters would be:

$$\text{Number of filters} = 2.7\sqrt{1.5}$$

Number of filters = 3

Flow

Next, the flow through one filter is calculated just as it was for one tank of the sedimentation basin:

$Q_{one} = Q_{total} / n$

$Q_{one} = (1.5 \text{ MGD})/3$

$Q_c = 0.5 \text{ MGD}$

So the flow through each of our three filters will be 0.5 MGD.

Surface Area

The required filter surface area is calculated using the formula below:

$A = Q_{one} / F.R.$

Where:

A = filter surface area, ft^2

Q_c = flow into one filter, gpm

F.R. = filtration rate, gal/min-ft^2

We will use a filtration rate of 4 gal/min-$ft.^2$ We will also have to convert from gpm to MGD.

The calculations for our example are shown below:

$A = 500{,}000$ gal/day × (1 day/1440 minutes)/4 gal/min-ft^2

$A = 87\ ft^2$

Tank Length

Since the filter tank is a square, the length of the tank can be calculated with the following simple formula:

$L = \sqrt{A}$

Where:

L = Length, ft

A = Surface area, ft^2

In the case of our example, the length of one tank is calculated as follows:

$L = \sqrt{87}$

L = 9.3 ft

This is the final calculation required for the design of the filter.

Clearwell Volume (Tank of backwash)

The volume of the clear-well must be sufficient to provide backwash water for each filter. First we calculate the total filter area:

Total filter area = A × (Number of filters)

For our example, the total filter area is:

Total filter area = 87 ft^2 × 3

Total filter area = 261 ft^2

Then we calculate the volume of the clearwell as follows:

V = (Backwash period) (Total filter area) (Filter rise rate)

We will assume a 5 minute backwash period and filter rise rate of 30 in/min. So, for our example, the volume of the clearwell would becalculated as follows:

$V = (5 \text{ min}) (261 \text{ ft}^2) (30 \text{ in/min}) (1 \text{ ft} / 12 \text{ in})$

$V = 3,263 \text{ ft}^3$

You will notice that we translated from inches to feet.

So, for our plant, we need three filters, each with a surface area of 87 ft^2 and a length of 9.3 ft. In order to accommodate backwashing all three filters at once, the clear well volume should be 3,263 $ft.^3$

References

Greiner, A. D., Timmons, M. B., 1998. Evaluation of the nitrification rates of microbead and trickling filters in an intensive recirculating tilapia production facility. Aquacultural Engineering, 18(3), pp 189 – 200.

Kamstra, A., Van der Heul, J.W. and Nijhof, M., 1998. Performance and optimization of trickling filters on eel farms. Aquacultural Engineering, 17(3), pp 175-192.

Saucier, B., Chen, S., and Zhu, S., 2000. Nitrification Potential and Oxygen Limitation in Biofilters. Presented at the Third International Conference on Recirulating Aquaculture July 2000.

Timmons, M.B. and Losordo, T. M., 1994. Aquaculture Water reuse Systems: Engineering Design and Management. Elsevier Science B.V.

Zhu, S. and Chen, S. 1999. An experimental study on nitrification biofilm performances using a series reactor system. Aquacultural Engineering, 20(4), pp. 245 – 259.

Wheaton, F. W., 1977. Aquacultural Engineering. John Willey & Sons Inc., USA.

5

Harvesting Equipments

5.1 Mechanized Harvesting in Fish Culture

Harvesting is a complex operation that can be divided into three major parts, as follows:

- Driving and concentrating the fish stock to the catch basin;
- Lifting the fish out of the catch basin and placing them on a sorting table;
- Sorting the fish and loading them on the transporting vehicle.

5.2 Herding operations

- Mechanical Driving
- Hydraulic Driving
- Electrical Driving

The driving and concentration of fish can be accomplished by three different ways, or by a combination of these, as follows:

- Mechanical driving by different nets;
- Hydraulic driving, i.e., draining the fish together with the water;
- Electrical driving (by electric grid).

5.2.1 Mechanical Driving

The mechanical driving of the fish by seines is the most commonly used method. When the fish pond is large, the netting should be mechanized because of the high pulling force. In some cases, conventional motorboats, tractors or trucks can be applied for pulling the net, but special winches have been developed for this purpose.

5.2.2 Hydraulic Driving

During hydraulic driving the fish are carried from the pond into the catch basin by water current as the water is draining from the pond. The pond and the drainage canal should be drained completely. The hydraulic drive is sometimes combined with mechanical or electrical drives. Open drainage canals are proposed to be built in these systems because the fish have a tendency to remain inside the dark underground pipe, even if the conditions are not good there.

5.2.3 Electrical Driving

Electric grids or electrified seines can also be used for driving, where the electric field forces the fish to swim towards the catch basin. Since some species of fish have a habit of swimming towards fresh inflow water, these fish species can be attracted to the catch basin by ensuring oxygen rich inflow water there. This can be combined with the above-mentioned driving methods.

5.3 Catching operations

- Mechanical Fish Lifting
- Hydraulic Fish Lifting

Lifting the fish out of the catch basin is a critical process because the fish are separated from their natural living medium: therefore, gentle handling is needed. Moreover, this process has to be intermeshed with the driving of fish into the catch basin and with sorting and transporting. Different lifting devices have been developed for fish harvesting, but the two major groups are mechanical and hydraulic devices.

Mechanical Fish Lifting

Mechanical fish lifting can be classified as follows:

- Devices with intermittent operation mechanized dip net bucket with alternate movement
- Devices with continuous operation conveyor belt screw bucket with rotating movement.

Mechanized dip nets are usually a mobile type of equipment and can be used widely, but two or three men are needed for the operation. The fish can be lifted out of the catch basin by a mechanized bucket elevator. One man can operate this device with a control panel but its installation is fairly complicated. It can only be installed in a well-built concrete structure. During harvesting, the fish may have contact with hard solid surfaces and one should account for some injuries.

Ordinary agricultural conveyor belts can be used for fish lifting as well, but specially designed conveyor belts have also been developed with soft rubber belts and covers. These belts are mobile equipment and can be widely used, especially when the fish have to be lifted up to a fairly high elevation. The lower hopper of the belts can be filled by hand or by some device like a mechanized dip net, air lift pump, *etc*.

Special fish screws have been developed recently for lifting fish. The internal spiral ribbon and the pipe housing revolve together lifting up the water in the space between the spiral ribbon and the housing. The applicability of the fish screws is the same as that of fish conveyor belts. However, the screw makes for more careful fish handling since the fish are lifted in water. Some other types of fish lifts with continuous operation have also been constructed where buckets are fixed to an endless chain or when a big rotating wheel with chambers is lifting the fish (Figure 5.1). However, these are not so widely used, mostly because of their fairly complicated installation and operation.

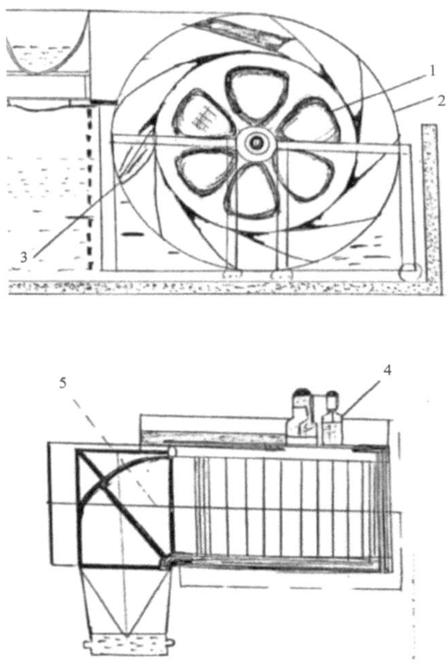

1. Wheel
2. Chamber
3. Air releasing device
4. Driving mechanism with electromotor
5. Discharge canal

Fig. 5.1: Continuous operation mechanized fish lifts with chambered wheel

Hydraulic Fish Lifting

The working principle of hydraulic fish lifting is basically different from that of mechanical lifting, as it is the water which is lifted directly and not the fish. Hydraulic fish lifting devices are usually water pumps which have been modified according to the special needs of fish lifting.

Hydraulic fish lifting devices or fish pumps can be divided into three groups, as follows:

- Centrifugal fish pumps
- Air lift pumps
- Vacuum pumps

Centrifugal fish pumps were first used in marine fishing vessels and, later on, in pond fish farming. These centrifugal pumps caused serious damage on 2-5 percent of the harvested fish by their impellers. Special impellers were later developed and the damage to the fish decreased. Recently, well designed centrifugal fish pumps became available and are commonly used, mostly in trout farms. However, the size of the fish is limited when using a centrifugal fish pump.

Air lift pumps are used in water management for delivery of water containing solid particles. The principle was adapted successfully for fish lifting, and special air lift pumps have been designed. Even big fish can be lifted without any damage, as there is no moving part inside the delivery pipe, only air bubbles.

The operation of Vacuum pumpsis intermittent and they can, therefore, be used most efficiently in small units (cages) where the whole stock can be sucked into the container during one cycle. These pumps can also be used for mud removal.

Fish Sorting

The harvested fish stock can be sorted according to size and, in polyculture, according to species. When the fish stock is sorted into species it has to be done by hand and the mechanization is restricted to transporting the fish to the feeding table. Usually, a horizontal conveyor belt is built into the table that conveys the dominant fish of the polyculture, and only the other species have to be touched by hand and placed aside.

The fish can be sorted into size by different screens. The screen can have several pairs of ribs with an increasing distance from top to bottom. The distance between two ribs is adjustable according to the grading requirement. There are graders where, instead of having fixed ribs, revolving rollers are used.

In order to achieve more accurate grading, there are graders where the fish does not merely slide down on the slope of the table by gravity but is led by an electric motor-driven endless rubber belt with soft rubber fingers above the slot.

Different types of scales are available to weigh the fish, but the four major types are described below:

- Steelyard
- Spring balance
- Tipping balance
- Displacement type of balance

Figure 5.2 shows displacement type of balance. A mobile fish conveyor belt or fish screw is the most efficient method of loading the fish from the sorting table to the transporting vehicle. The mechanization of fish harvesting does not necessarily mean total mechanization. In some cases, only one phase of the operation is mechanized but there are farms where complex mechanized harvesting lines have been installed. Diagrams of three of these complex mechanized harvesting systems are shown in Figures 5.3, 5.4 and 5.5.

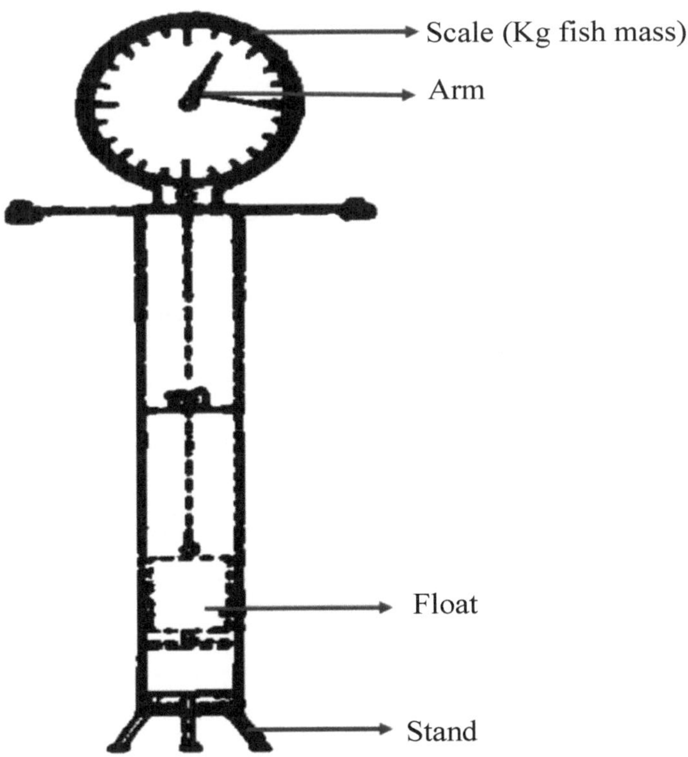

Fig. 5.2: Water displacement type fish weigher (GDR)

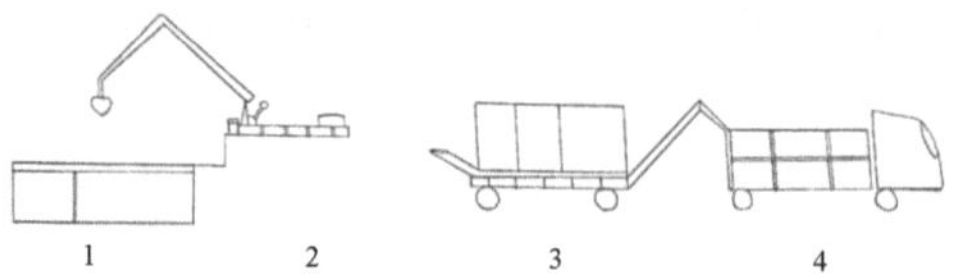

1. Concrete catch basin with fish driving screen
2. Mechanized dip net
3. Sorting platform
4. Transport vehicle

Fig. 5.3: Mobile mechanized fish harvesting system

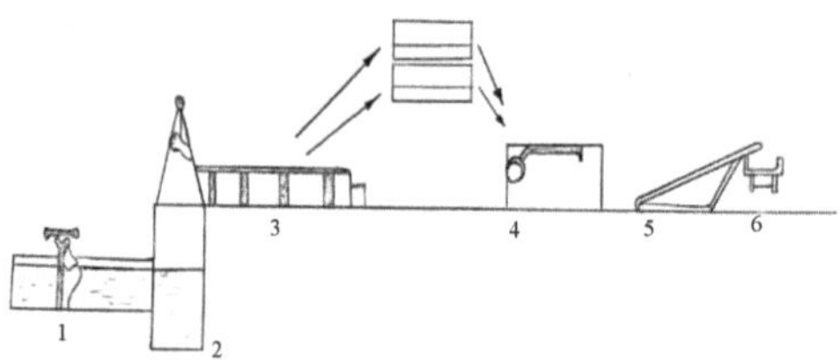

1. Concrete catch basin with fish driving screen
2. Bucket elevator
3. Sorting table
4. Steelyard
5. Belt conveyor
6. Transport vehicle

Fig. 5.4. Stationary mechanized fish harvesting system

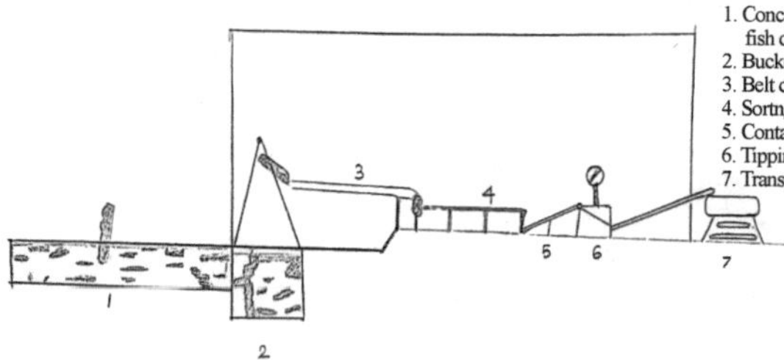

1. Concrete catch basin with fish driving screen
2. Bucket elevator
3. Belt conveyor
4. Sortng table
5. Container with belt conveyor
6. Tipping scales
7. Transport Vehicle

Fig. 5.5. Central fish harvesting place with indoor sorting

References

Fonarev, A.L., 1977, Studies on hydraulic transportation of fish. Moskva, Pischevaja Promyshlennost

Greenland, D.C., 1974, Recent development in harvesting, grading, loading and handling pond raised catfish. Trans. Am. Soc. Agric. Eng., 17(1)

Kuli, B. and I. Csávás, 1976, A tógazdaságihaltermelésgépesitése (Mechanization of pond fish production). Budapest, AGROINFORM

Kuli,B.,1976,Különbözögépesitettlehalászásirendszerekösszehasonlitóvizsgálata(Comparative trial of different mechanized harvesting methods). Gödöllö, MÉMMüszakiIntézet

Tjuktjaeva, J.S. and V. Ja Lysenko, 1974, Handbook on mechanization of processes of pond fish culture. Moskva, PischevajaPromyshlennost

Torban, S.S., 1977, Mechanization of processes of industrial fisheries. Moskva, Pischevaja Promyshlennost

Vibert, R. (ed.). 1967, Fishing with electricity: its application to biology and management. London, Fishing News (Books) Ltd., 276 p.

6

Material Handling Equipment

Material handling includes a number of operations that can be executed either by hand (manual) or by mechanical means or devices **to convey material** and **to reduce the human drudgery.**

The most common types of mechanical devices for grain handling are;

1. Belt conveyor
2. Bucket elevator
3. Screw conveyor
4. Chain Conveyor
5. Pneumatic conveyor

Selection of material handling machines and conveyors

The selection of proper conveying system is important for ease in operation and getting desired capacity for a particular product. Principles based on which the material handling equipment is selected:

- Based on the characteristics of the products being conveyed
- Working and climatic conditions.
- The capacity of conveying
- In a conveying system possibility of use of gravity.
- The capacity of handling / conveying equipment should match with the capacity of processing unit or units.
- Spillage of conveyed products should be avoided.
- Pollution of the environment due to noise or dust by the conveying system should also be avoided.

Belt conveyors

A belt conveyor is an endless belt operating between two pulleys with its load supported on idlers. The belt may be flat for transporting bagged material or V-shaped. The belt conveyor consists of a belt, drive mechanism and end pulleys, idlers and loading and discharge devices (Figure 6.1).

Fig. 6.1: Belt Conveyor

On the belt conveyor baggage/ product lie still on the surface of belt and there is no relative motion between the product and belt. This results in generally no damage to material. Belt can be run at higher speeds, so, large carrying capacities are possible. Horizontally the material can be transported to longer distance. The initial cost of belt conveyor is **high for short distances,** but for longer distances the initial cost of belt conveying system is low. The first step in the design of a belt conveyor with a specified conveying capacity is to determine the speed and width of the belt.

The belt speed should be selected to minimise product spillage or removal of fines due to velocity of the belt. For transportation of grains, the belt speed should not increase 3.5 m/s. Generally, for grain conveying, belt speed of 2.5 to 2.8 m/s is recommended. The selection of belt width will depend upon the capacity requirement, speed of operation, angle of inclination of belt conveyor, trough angle and depth.

The capacity of belt conveyor can be calculated as:

$$Capacity\ (m^2/h) = Area\ of\ cross\ section\ (m^2) \times Belt\ speed\ (m/min) \times 60$$

Belt conveyor idlers: The efficiency of belt conveyor is largely dependent on idlers. For higher efficiency of belt conveying systems, the idlers must be accurately made and provide a rigid framework. This will maintain a permanent, well-balanced smooth-running alignment.

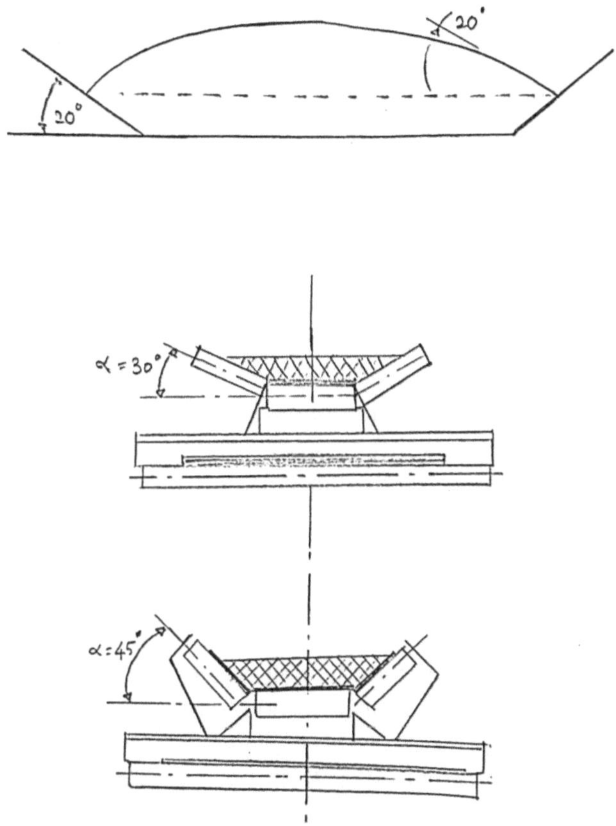

Fig. 6.2: Different troughing configurations

There are three kinds of belt carrying idlers which are used in handling of bulk materials (Figure 6.2). The type of idlers affects the cross-sectional load on the belt.

1. The **flat belt idlers** are used for granular materials having an angle of repose of not less than 35°.
2. **Troughing idlers with 20°** trough is used for conveying all kinds of bulk materials.
3. **Troughing idlers with 35° and 45° trough angle** is mainly used for transportation of small particle lightweight materials like grain, cotton seed etc. It is also used for carrying heavier, medium size lumps like crushed stones.

Idler Spacing

The spacing between the idlers influences the retention of correct troughing. The incorrect idler spacing may result in belt undulation. The pitch of idlers is determined by the idler load rating or the carrying capacity of each idler, on the sag of the belt between the idlers, belt tension and belt speed. As a token, the space between the successive idlers should be approximately equal to the width of belt. The spacing should not exceed 1·2 meters.

Belt tension

The tension developed at the drive pulley in transmitting the required power to move the loaded belt is known as effective tension. The effective tension is the sum of tension to move the empty belt, the tension to move the load horizontally and the tension to lift the material. The effective tension is related with the power required to move the belt and belt speed in the following manner.

$$Effective\,Tension, T = \frac{Power\ in\ kW}{belt\ Speed, m/s}$$

Grains are mostly discharged from the belt conveyor over the end pulley or at any other point along the conveyor by a scraper plough or a throw-off carriage known as a tripper. While leaving the belt over the end pulley, product flow will describe the path of a parabola. Belt conveyors can discharge grains at various locations by means of a movable tripper (Figure 6.3). Trippers are available as hand propelled, self-propelled or automatic.

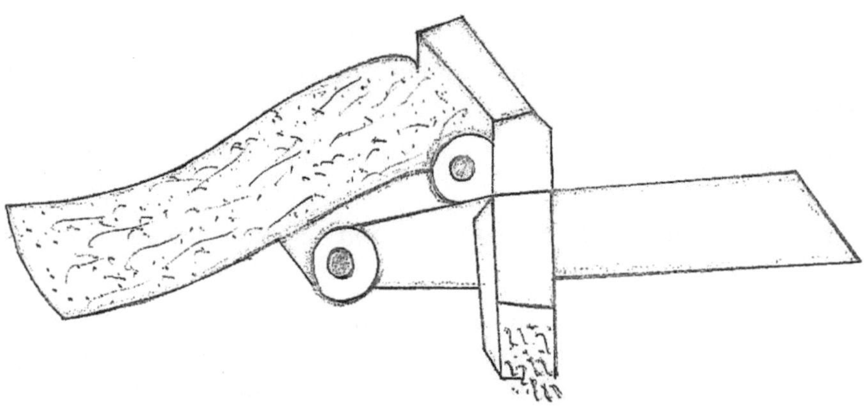

Fig. 6.3: Tripper for discharge of materials

Bucket Elevator

A bucket elevator consists of buckets attached to a chain or belt that revolves around two pulleys one at top and the other at bottom. The vertical lift of the elevator may range between few metres to more than 50 m. **Capacities of bucket elevators may vary from 2 to 1000 t/hr.** Bucket elevators are broadly classified into two general types, (1) spaced bucket elevators and (2) continuous bucket elevators.

The spaced bucket elevators are further classified as,

1. Centrifugal discharge elevators,
2. Positive-discharge elevators,
3. Marine leg elevators and
4. High-speed elevators.

The continuous bucket elevators are classified as

1. Super capacity bucket elevators and
2. Internal-discharge bucket elevators.

The spaced bucket centrifugal discharge type (Figure 6.4) is most commonly used for elevating the grains. The bucket elevator is a very efficient device for the vertical conveyance of bulk grains. Bucket elevators with belts are employed in food industries for vertical conveyance of grains, derivatives and flours. Bucket elevators are usually mounted at a fixed location, but they can also be mounted in a mobile frame. Bucket elevators have high capacities and it is a fairly cheap means of vertical conveyance.

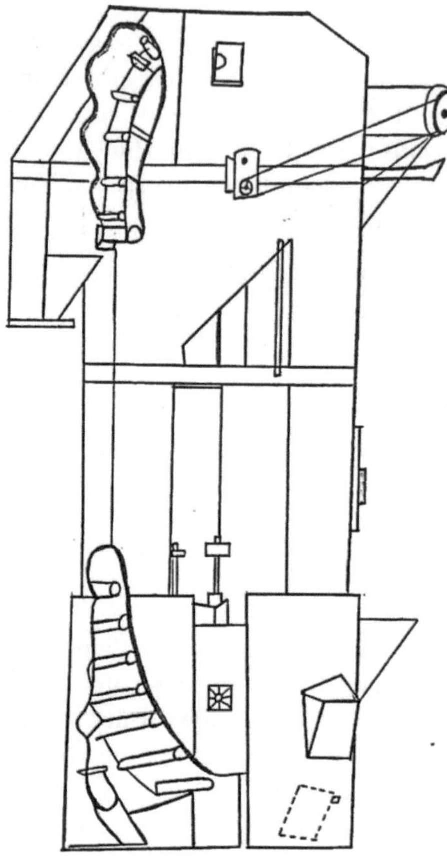

Fig. 6.4: Bucket Elevator

It requires limited horizontal space and the operation of conveying is enclosed in housing, thus it is dust free and fairly quite. The bucket elevator has limited wear problem since the product is enclosed in buckets.

The buckets are enclosed in a single housing called leg, or two legs may be used. The return leg may be located at some distance from the elevator leg. The boot can be loaded from the front or back or both.

The product flow is discharged either by means of gravity or centrifugal force. The bucket elevator's capacity mainly depends on bucket size, conveying speed, bucket design and spacing, the way of loading and unloading, the bucket and the characteristic of bulk material. Belt speed is the first critical factor to consider. Bucket elevators with a belt carrier can be used at fairly high speeds of 2.5 to 4 m/ s.

The bucket elevator's capacity may be calculated by the following equation.

Elevator capacity (m^3/h) = Bucket capacity (m^3) × No. of buckets per meter of belt× belt, speed (m/min) × 60

The main parts of a bucket elevator are,

1. Elevator head section,
2. Elevator boot section,
3. Elevator legs,
4. Belts for bucket elevator and
5. Buckets.

Head Section

The discharge side of the high-speed conventional bucket elevator's head should be shaped so that material thrown from the buckets may not deflect into the down leg. When the product is not thrown well clear of the buckets into the discharge chute, it will fall in the down leg. This is called as "back logging".

Boot Section

Bucket elevator boots should be of bolted assembly to allow for proper maintenance and replacement of pulley, shaft and other accessories.

Elevator Legs

The up and down moving string of buckets in bucket elevators are enclosed in elevator legs. The elevator legs limit the emission of dust. These legs are constructed as all welded, bolted or riveted units.

Elevator Belts

The bucket elevator belt has no support between the drive and the return pulleys, therefore, cross stiffness of belt is very important. The stretch limitation of the conveyor belts is also very important. The total stretch of the belt under maximum load should not exceed 1-2 % of the belt length.

A belt is as strong as its weakest point, so a through connection of the belt ends is important. Several types of splicing are possible. The type of belt splice depends on the thickness of the belt and the severity of service. For belts of five ply thickness or less, the bolted clamp joint, the lap joint, or the butt-strap joint may be used.

Buckets

As per the requirements, buckets are made of different materials and come in various shapes and sizes. The shape of the bucket is very important for filling and discharge. The top angle is generally taken as 80° while the bottom angle is between 20-30°.

Drive mechanism

The drive mechanism of a bucket elevator is located near the elevator head. At the elevator head, the belt is turned around the drive pulley. Drive motor with gear box and couplings are mounted on a rigid and separate frame.

HP requirement

The theoretical horsepower requirement for the bucket elevator can be calculated by the following equation.

Power (hp) = QHF/4562

where,

Q = capacity of bucket elevator, kg/min

H = lift of elevator, m

F = factor; 1.5 for elevators loaded on the up side and 1.2 for elevators loaded on the bottom side

The theoretical horsepower should be increased 10-15% to provide for friction and power requirements for loading, power transmission and drive losses.

Screw Conveyor

The screw conveyor consists of a tubular or U-shaped trough in which a shaft with spiral screw revolves (Figure 6.5). The screw shaft is supported hanger bearings at ends. The rotation of screw pushes the grain along the trough. The screw conveyor is used in grain handling facilities, animal feed industries and other installations for conveying of products generally for short distances. Screw conveyor requires relatively high power and is more susceptible to wear than other types of conveyors. The pitch of a standard screw which is the distance from the centre of one thread to the centre of the next thread is equal to its diameter. For example, a 10 cm diameter screw has a pitch of 10 cm.

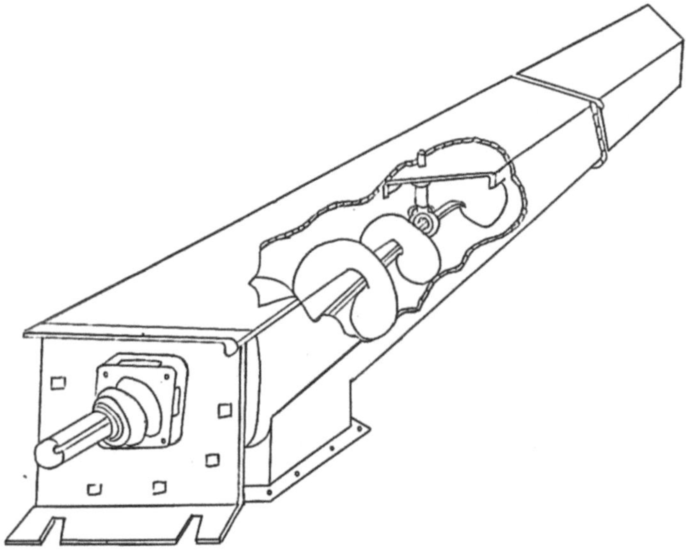

Fig. 6.5: Screw Conveyor

As the screw conveyor's driving mechanism is simpler, and no tensioning device is required, the initial cost of the conveyor is lower than any other conveyor with the same length and capacity. The main parts of a screw conveyor are, screw blade, screw shaft, coupling, trough, cover, inlet and outlet gates, bearings and drive mechanism.

The screw conveyor is generally used to move grains horizontally. However, it can also be used at any angle up to 90° from the horizontal, but the capacity correspondingly reduced as per the inclination of conveyance.

The screw basically consists of a shaft and the screw blade or flight. The flight is a continuous one-piece helix shaped from a flat strip of steel welded onto the shaft. The screw shaft is usually a joint less tube with thick sides and a high tensile strength to reduce the weight. The thickness of the steel strip helix decreases from the inner edge to the outer edge. Troughs of screw conveyor have different shapes. Most common is U shaped trough. In an enlarged or flared trough, the side walls become wider at the top. This type of trough is usually used for conveying non-easy flowing materials which may have lumps. The tubular trough is completely closed with circular cross-section and mostly used for conveying materials at inclination or for vertical lift.

If the screw conveyor is used to convey different materials, mixing of products is possible. The capacity of screw conveyor is influenced by the screw diameter, inclination of the screw blade, speed of the blade, shaft diameter and cross-section of loading. The theoretical conveyance capacity of the screw conveyor can be given by the following equation.

Capacity Q, m^3/hr= 47.2 $(D^2- d^2)$ X P x n

where,

D = screw diameter, m

d = shaft diameter, m

p = pitch, m

n = rpm

The theoretical capacity is more than the actual capacity because of screw housing clearance and the loading factor

The power requirement of screw conveyors for horizontal operation is determined by the following equation.

Horse power = QLWF/4560

where, Q = conveyor capacity, m^3/hr

L = conveyor length, m

W = bulk material weight, kg/m^3

F = material factor (for paddy 0.4)

If the calculated horsepower is less than 1.0, double the value.

horse power = 1 to 2 multiply the value by 1.5

horse power = 2 to 4 multiply the value by 1.25

horse power= 4 to 5 multiply the value by 1.1

For horsepower values of more than 5, no correction is required.

Pneumatic Conveyor

The pneumatic conveyor moves granular materials in a closed duct by **a high velocity air stream**. Pneumatic conveying is a continuous and flexible transportation method. The material is carried in pipelines either by suction or blowing pressure of air stream. The granular materials because of high air pressure are conveyed in dispersed condition. For dispersion of bulk material, air velocities in the range of 15-30 m/s is necessary.

The pneumatic conveying system needs a source of air blowing or suction, means of feeding the product into the conveyor, ducts and a cyclone or receiving hopper for collection of products. There are three basic systems of pneumatic conveying. These are pressure or blowing system, suction or vacuum system, and combined push-pull or suck blow system. In blowing or positive pressure systems, the product is conveyed by using air pressures greater than the atmospheric pressure.

The selection of air mover is the most important aspect of the design of a pneumatic conveying system. In design, the two factors, (1) supply air pressure and (2) the volumetric flow rate of air should be considered.

For separation of product particles from air, air-product separators are used. Cyclones are mostly used to collect the particles. Cyclone is a device which removes the bulk of the product particle from the conveying air stream by centrifugal force. In some cyclone, a fabric filter is attached to remove residual dust and fine product particles from the air stream.

Limitations of Pneumatic Conveying

1. Erosion of solid surfaces and equipment surfaces by solid particles with conveying air stream.
2. In case of bends or misaligned sections, the erosion problem becomes severe.
3. Chances of repeated impacts between the particles and the solid surfaces are high. Due to such impacts, product degradation results.

7

Electronic Equipments and Devices Used in Aquaculture

7.1 Measuring Dissolved Oxygen

The oxygen content may be measured either chemically or electronically. The normal chemical method is the so-called Winkler method which is a titration method, consisting of adding certain chemicals to the water and observing a colour change which will be directly related to the oxygen concentration.

When measuring oxygen *concentration* electronically, a sensor or probe is used. This can be constructed using a positively charged conductor (anode) and a negatively charged conductor (cathode) separated by an insulator. In one design, an electrolyte is located around the electrodes (anode and cathode). A special membrane covers the electrolyte, keeping it in place and protecting it. The membrane is permeable to oxygen molecules. As there is a positive and a negative electrode, electrolysis will occur, and electrons will pass between the two electrodes. The magnitude of the electron transport is affected by the amount of *oxygen* in the electrolyte; this again depends on the amount of oxygen transported through the membrane, which is related to the amount of dissolved oxygen in the water. The typical DO meter is shown in Figure 7.1.

Fig. 7.1. DO meter

It is important to calibrate this type of instrument accurately. For some instruments it is necessary to calibrate for air pressure and temperature, while in others this is integrated. The electrolyte and the membrane or the complete sensor, must be changed at defined intervals. Since the membrane is very thin it is easy to break, care must be taken when handling it. Ageing of the membrane can be detected, because it becomes difficult to get stable results and the values fluctuate constantly. The membrane is also exposed to fouling, and must regularly be visually inspected and cleaned. If fouling occurs the oxygen values will normally drop in relation to the correct values.

7.2 Measuring pH

pH is a measurement of the degree of acidity or alkalinity of water. It is measured on a scale of 0 to 14. General health and wellbeing of the culture stock can be attributed to the pH level. Every cultured species has a preferred pH range in which it will grow best and breed. A pH level near to their native environment is the best. Maintaining a correct pH is a constant concern for the aquaculturist. pH adjustments require several careful dosages over a period of time instead of a single large dosage. Large pH fluctuations can cause shock to all species and result in heavy casualties. A well-maintained pH level will result in less stress and thus reducing sickness and deaths.

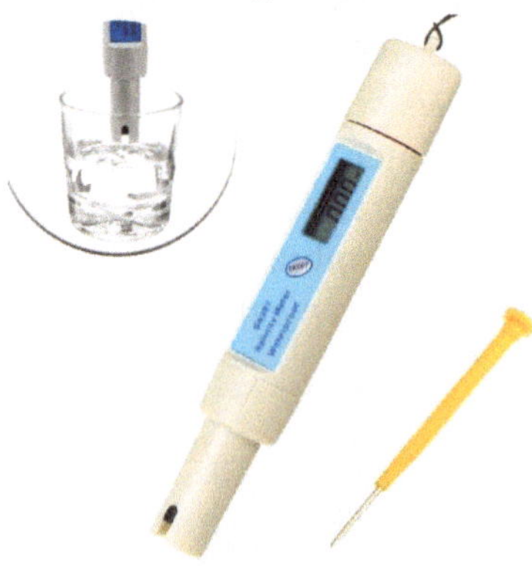

Fig. 7.2. pH meter

A pH meter (Figure 7.1) is constructed on the same principle as the oxygen meter. Here the probe is constructed with two electrodes, one for measuring the concentration of H+ and the other as a reference electrode. In this case the pH electrode consists of a membrane permeable to hydrogen ions. Here also the membrane creates a cell that encloses the electrolyte. Between the electrodes a current will pass that depends on the concentration of H+. The voltage between the pH electrode and the reference electrode is then measured, and the pH calculated.

Instruments for measuring pH must be calibrated before use. This is performed by placing the probe in different solutions of known pH and then adjusting the instrument. The pH probe is compact, in contrast to the O_2 probe; therefore it must be completely replaced if it ceases to function. A disadvantage of the pH probe is its limited duration, normally from 3 months up to 1 year in special cases. Practical experience with the use of pH meters in fish farming shows that maintenance is of extreme importance. Both calibration and changing of electrodes must be done regularly to achieve reli-able measurements.

7.3 Measuring Conductivity and Salinity

Water and soil salinity are measured by passing an electric current between the two electrodes of a salinity meter in a sample of soil or water. The electrical conductivity or EC of a soil or water sample is influenced by the concentration and composition of dissolved salts. Salts increase the ability of a solution to conduct an electrical current, so a high EC value indicates a high salinity level.

Electrical conductivity (EC) is also a term used to describe a measurement unit of salinity. Conductivity is a measure of the ability of water to conduct an electric current. In fish farms this is important in order to evaluate the ability of the water to inhibit pH fluctuations, i.e. the buffering capacity. In seawater, Na^+ and Cl^- ions dominate and here the salinity meter (Figure 7.3) is used to measure the salinity.

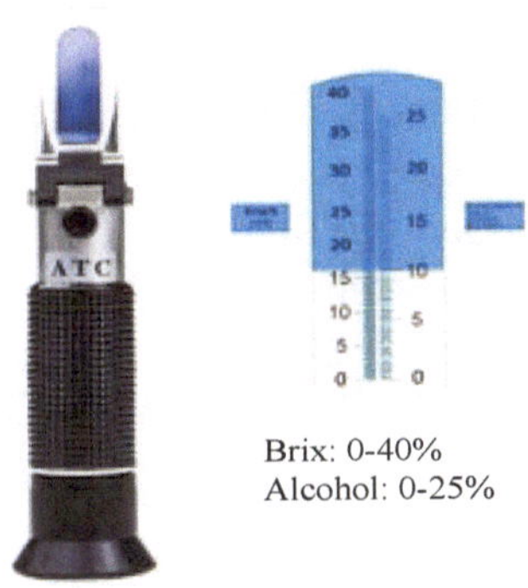

Fig. 7.3. Salinity meter

The probe consists of two electrodes and is lowered into the water. A small electric potential (voltage) is applied across the electrodes. An electric current will occur between the electrodes, the size of which depends on the ion concentration in the water. To prevent the establishment of a layer on the electrodes which affects the current, it is necessary to use an alternating current as pre-voltage.

Conductivity is affected by temperature, so it is important to compensate for this parameter when taking measurements. Each instrument should have a special table setting out the effect of temperature on the conductivity. Advanced instruments incorporate automatic temperature compensation.

7.4 Measuring Water Velocity

The water velocity is proportional to water flow. If the water velocity and the inter-nal diameter of the pipe are known, it is easy to calculate the water flow.

A simple way to measure the water velocity in a pipe is to use a propeller, paddle wheel or turbine Propellers with a variety of designs are used. The working principle is that the water will move the propeller and the rotational velocity of the propeller will reflect the water velocity in the pipe. Either the propeller can be installed in an existing pipe system, known as an inset meter, or it can be a completely separate system in which the propeller and instruments are connected, forming a complete unit adapted to the pipe. Propeller systems are simple and inexpensive, but one of the disadvantages is that the head loss will increase.

However, the main disadvantage is that the propeller will be exposed to fouling because it is within the water flow. Normally, fouling will reduce the measured velocity compared with the actual velocity. Correct maintenance is therefore important in such a system. A further disadvantage is wear of the continuously moving propeller.

An electromagnetic flow meter can also be used to measure water velocity. The principle utilized here is that an electromagnetic field changes when the water velocity varies. A conducting fluid, such as water, in a magnetic field will induce an electric voltage, which correlates with the water velocity. The water must have some electrical conductivity if this type of instrument is to function; in fish farming this will always be the case.

Ultrasound waves may also be used for measuring the water flow, as in an ultrasonic flow meter. Two combined ultrasound source and receiver units are placed, one on each side of the pipe and crossed diagonally. Sound waves are sent from one side and travel with the water flow; from the opposite side of the pipe another sound wave is sent against direction of the water flow. The times taken for the sound waves to reach the receivers are compared; the water velocity can be calculated because the water flow deflects the sound waves and alters their velocity. A fixed unidirectional ultrasonic system can also be used which calculates the time difference during which the sound waves travel. The advantage of this system is that the instrument may be attached to the exterior of existing pipes; if the type of pipe is known the water flow inside can be found. Flow meters utilizing this principle will not give any head loss and are very accurate; however, they are quite expensive. The instruments may also be affected by air bubbles or high particle concentrations.

The various types of flow meter have different accuracies, and this must be checked before selection. Generally, greater accuracy incurs greater expenditure, so the meter must be fit for purpose.

7.5 Measuring Water Level

Measuring the water level is necessary at various places to avoid overflows and water shortages. This can be carried out in a head tank or directly in the individual fish tanks. Water level sensors are normally digital, sending a signal if the level of the water is below or above defined values. The system may also be analogue; this generates a signal that is proportional to the water level.

Different types of electronic floats may be used for controlling the water level. The float will lie on the surface and change position if the water level varies; this variation can be utilized to control the level. When this happens a signal will be sent and action taken, for instance switching the water supply on and off. A commonly used float is the level rocking sensor which floats on the surface; if the water level decreases it will gradually become more upright. The sensor is located in theend of a cable and when it is in the floating position there may or may not be contact between the two conductors located inside the float. If the sensor is hanging vertically an automatic switch will take the opposite position (opened or closed as the case may be) to make or break the electric circuit. This can then be used to control the water level. On submerged pumps this is a common method for starting and stopping the pumping, depending on the water level

7.6 Seechi disk

A Secchi disk (Figure 7.4) is a standard tool used to measure water transparency in fishponds to determine the amount of plankton. Plankton causes the water to turn green and serves as natural food for the fish. The disk, which consists of a circular plate painted with alternating black and white quadrants, is typically mounted on a numerically graduated string and lowered into the water. The depth at which it is no longer possible to tell the difference between the black and white parts of the disk is taken as a measure of the transparency of the water.

Fig. 7.4. Secchi Disk

A Secchi Disk also used to measure turbidity. The Secchi disk is lowered into the water until it is no longer visible, and that depth is measured. Secchi depth values that are high indicate clearer water, and low Secchi depths indicate high turbidity.

7.7 Measuring Turbidity

Turbidity is the measure of suspended solids in the water. When the water is clear, there is low turbidity; when the water is cloudy, there is high turbidity, which allows less light to penetrate through the water. Light is required for seabed grasses to flourish.

What causes high turbidity? Run-off from the land due to stormwater, shoreline erosion, excessive algal blooms, and pollution all contribute to high turbidity levels.

7.8 Blower

Aeration in aquaculture is becoming more and more important with the increase in stocking density and use of new age farming methods like RAS, biofloc.

Blower is a device it is used to blow the oxygen into the water. Centrifugal Air Blowers are highly effective in many manufacturing processes, such as where cooling, cleaning, blow-off, drying and moisture regulation is needed. The key parts of Centrifugal Air Blowers are their impellers, or rotors. These are integral to how the blower increases air pressure and creates air flow. Air enters the centre of the impellers, and then is divided between the impeller blades. This increases the volume of the air stream, and increases the speed of the air using centrifugal force. The high-velocity air that is accelerated as a result is then contained in the blower housing, creating pressure.

The rotating impellers create, essentially, a vacuum. This pulls air through the blades, pushing it out the other side. This means that whilst air velocity - the measure of the distance and speed air is travelling (usually defined in meters per second) - is fixed, the mass flow rate - the mass of air passing a point in the system (usually defined in kilograms/pounds per second) - is not.

So, this is how Centrifugal Air Blowers can use lower flows to create higher velocity, and so operate at a low-pressure ratio. This makes Centrifugal Blowers lower cost to run than compressed air systems.

Centrifugal Air Blowers (Figure 7.5) use the kinetic energy of their impellers to increase the volume of an air stream. In short, they work by transferring energy between their impellers and the air.

Maintaining low pressure is key to how Centrifugal Air Blowers work. Blower systems are tested to ensure this is the case using two different pressure measurements - local pressure - the pressure at a specific point - and cumulative pressure - the total pressure from the entire system.

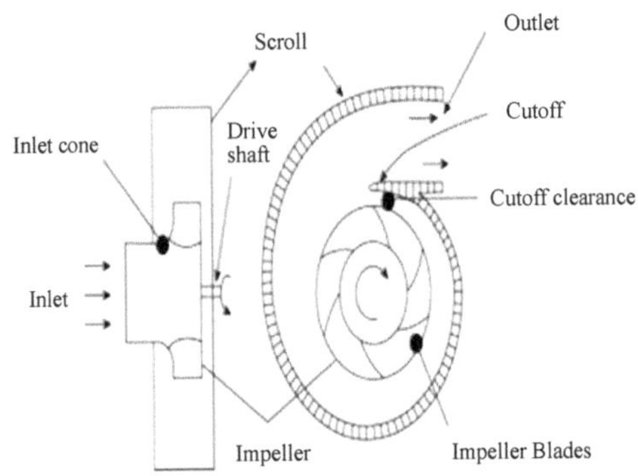

Fig. 7.5. Centrifugal air blowerRud magna adiat. Raesed eum iriure euis nulla faccum-mod tate tie vulluptat, cons niamet et lore consequi er sim quisi.